STUDY GUIDE

CLARK E. ADAMS

ENVIRONMENTAL SCIENCE

TOWARD A SUSTAINABLE FUTURE

EIGHTH EDITION

RICHARD T. WRIGHT
BERNARD J. NEBEL

Upper Saddle River, NJ 07458

Executive Editor: Dan Kaveney
Assistant Managing Editior: Dinah Thong
Associate Editor: Amanda Griffith
Executive Managing Editor: Kathleen Schiaparelli
Production Editor: Natasha Wolfe
Supplement Cover Manager: Paul Gourhan
Supplement Cover Designer: PM Workshop Inc.
Manufacturing Buyer: Lisa McDowell
Cover Photograph: Offshore oil platform-Ken Graham/Stone; Solar energy station-IFA Bilderteam/Leo De Wys; Modern Windmills-Library LTD/Leo De Wys; Windmill on farmstead-Greg Latza. PeopleScapes, Inc.

Upper Saddle River, NJ 07458

Printed in the United States of America

10 9 8 7 6 5 4 3 2 1

ISBN 0-13-091391-X

Prentice-Hall International (UK) Limited, London
Prentice-Hall of Australia Pty. Limited, Sydney
Prentice-Hall Canada, Inc., Toronto
Prentice-Hall Hispanoamericana, S.A., Mexico City
Prentice-Hall of India Private Limited, New Delhi
Pearson Education Asia Pte. Ltd., Singapore
Prentice-Hall of Japan, Inc., Tokyo
Editora Prentice-Hall do Brazil, Ltda., Rio de Janeiro

TABLE OF CONTENTS

TO THE STUDENT

HELPFUL HINTS ON HOW TO USE THIS STUDY GUIDE

I developed this study guide by playing the role of the student who would be using it. For example, you will note that the study guide topics are correlated with and in the same sequence as presented in the text. As you read each chapter of the text, you should easily find the answers to each question. However, this will not be the case if you are using an earlier edition of Environmental Science Towards a Sustainable Future. This study guide was developed specifically for the latest edition of this text.

The study questions and vocabulary drill were constructed to give you a directed system of learning, thus minimizing wasted effort. Your success on the self-tests should give you some indication of whether you have indeed arrived at one level of mastery of chapter content.

There are two ways you can use this study guide that will help you reinforce text content. One way is to first read an entire chapter and then attempt to answer as many of the study guide and vocabulary drill questions as possible. The correlations of study guide and textbook chapters and subsections should make it easy to look up answers. Any answers you need to look up may require another reading of this section of the text and which should also alert you to chapter content areas that need additional attention.

Another way to use this study guide is to answer each question as you read the text. First read the study guide question, then read the text until you find the answer to that particular question. You will find that using this system will, at times, require that you read several pages before finding the answer to a particular question. This system of learning works because it focuses your study on one issue at a time.

After you have attempted to answer all study guide and vocabulary drill questions, check your answers with the keys at the end of the chapter. Finally, answer the self-test questions and grade yourself using the self-test answer key at the end of the chapter.

Remember that incorrect responses to either the study guide, vocabulary drill, or self-test questions are as valuable indicators of your mastery of chapter content as are correct answers. The incorrect responses alert you to specific areas that need additional attention and thus focuses your study time.

Concentrated study of the information in the text and study guide will give you a comprehensive understanding of the pertinent ecological principles, how various human activities are in conflict with these principles, and how this conflict might be mitigated by redirecting human activities to be in concert with ecological principles. If you can address each issue on these three levels, then you should enjoy a high level of personal satisfaction and academic success.

CHAPTER 1

INTRODUCTION: SUSTAINABILITY, STEWARDSHIP, AND SOUND SCIENCE

1. Characterize how differently people in developing and industrialized countries are tied to their environment.

   a. developing countries: ______________________

   b. industrialized countries: ______________________

2. The historic residents of Easter Island learned that civilization collapses when:

   a. ______________________

   b. ______________________

   c. ______________________

3. What four steps need to be taken to prevent a global version of the collapse of civilization on Easter Island?

   a. ______________________

b. ______________________________

c. ______________________________

d. ______________________________

The Global Environmental Picture

4. What global trends represent views of the interactions of human and natural ecosystems?

   a. ____________________ b. ____________________

   c. ____________________ d. ____________________

Population Growth

5. Give the world population figures in:

   a. 1999 ____________________ b. 2200 ____________________

6. Given the rapid human population growth rate and expected global population size by 2050, discuss the growing conflict between "global quality of life" and "American lifestyle."

   ____

   ____

The Decline of Ecosystems

7. What are four indicators of the decline of vital ecosystems?

   a. __________.__________ b. ____________________

   c. ____________________ d. ____________________

Global Atmospheric Changes

8. Give one example of an air pollution problem with global dimensions.

   ____

9. The carbon dioxide level in the atmosphere has grown from ____________________ parts per million (ppm) in 1900 to ____________________ ppm today.

10. Explain the expected relationship between atmospheric carbon dioxide and the greenhouse effect.

    ____

    ____

Loss of Biodiversity

11. List two ways in which habitat alteration is taking place.

    a. ______________________ b. ______________________

12. What three factors are the greatest contributors to loss of biodiversity?

    a. ______________________ b. ______________________

    c. ______________________

13. Give three reasons why the loss of biodiversity is so critical.

    a. ____

    b. ____

    c. ____

Three Unifying Themes

14. List three interrelated themes that may be applicable to changing or giving direction to the interaction between human and natural ecosystems.

    a. ____

    b. ____

    c. ____

15. The current trends of human uses of the world's natural resources (is, is not) sustainable.

16. Imagine you are in the commercial fishing (e.g., tuna) business. Fisheries biologists have estimated that the fishery consists of 10K tuna, of which 8K are the surplus resulting from the reproducing component of the population. In order to maintain a sustainable yield, how many tuna should be harvested? ______________

17. Justify your answer to question 16.

    ____

    ____

18. List three characteristics of a sustainable society.

    a. ____

    b. ____

    c. ____

Sustainable Development

19. Apply the meaning of sustainable development to the disciplines of:

 a. economics: ______

 b. sociology: ______

 c. ecology: ______

20. List four dimensions to sustainable development.

 a. ______

 b. ______

 c. ______

 d. ______

Stewardship

21. Describe three ways in which the stewardship ethic is demonstrated.

 a. ______

 b. ______

 c. ______

Justice and Equity

22. What is the major problem being addressed by the environmental justice movement?

 ______

23. How is environmental racism demonstrated? ______

 ______

24. (True or False) The perpetrators of environmental racism come only from outside developing countries.

25. List four conditions (local and global) that contribute to poverty in developing countries?

 a. ______ b. ______

 c. ______ d. ______

26. (True or False) Failure of debtor nations to repay their loans will not hurt the economies of developed nations.

Environmentalism

27. Trace the conservation attitude of Americans during the following time periods by indicating whether the attitude was positive [+] or negative [-].

[ ] late nineteenth century
[ ] scientific era after World War I
[ ] depression years (1930 to 1936)
[ ] during World War II
[ ] after World War II until the 1960s
[ ] from the mid 1960s to 1980
[ ] 1980 to now

28 Identify one environmental problem that developed during the post-World War II era in the

a. air: ______________________________

b. waters: ______________________________

c. wildlife: ______________________________

29. What "cause-and-affect" relationship did Rachel Carson portray in her book Silent Spring?

____

30. Why does the position of citizens aligned with the environmental movement include wildlife?

____

31. Identify four environmental organizations that grew out of the environmental movement.

a. ____________________ b. ____________________

c. ____________________ d. ____________________

32. List four successes associated with the environmental movement.

a. ______________________________

b. ______________________________

c. ______________________________

d. ______________________________

33. What are the two major points of contention in the spotted owl debate?

a. ____________________ vs. b. ____________________

34. (True or False) Polluters represent one group of people.

35. Indicate whether the following statements are [+] or are not [-] examples of the policy recommendations of the President's Council on Sustainable Development.

[ ] restoration of biodiversity of terrestrial and aquatic ecosystems
[ ] identification of the real stakeholders in the process
[ ] enhancing agricultural productivity
[ ] coordination of stakeholders to achieve specific goals
[ ] habitat restoration
[ ] partnership development

Sound Science

36. Scientific information is based on repeated ________________.

37. The art of compiling scientific information is a series of steps called the ________________ method.

38. List the four basic assumptions that form the unquestioned foundation of the scientific method.

a. ________________________________

b. ________________________________

c. ________________________________

d. ________________________________

39. The scientific method begins with:

a. observations b. developing hypotheses c. experimentation d. formulation of theories

40. In science, the only observations accepted are those made with the basic five senses, namely ________________, ________________, ________________, ________________, and ________________.

41. Before it is accepted, an observation must be confirmed or ________________.

42. Indicate whether the following statements are [+] or are not [-] correct.

[ ] All observations are facts.
[ ] All facts are verified observations.

43. The purpose of a scientific experiment is to:

a. test a hypothesis b. prove that a hypothesis is correct c. confuse the public
d. use fancy instruments e. have fun at public expense

44. A hypothesis that cannot be tested is:

a. correct b. incorrect c. useful d. useless

45. To give definitive results, an experiment must involve a minimum of __________ groups.

46. The two groups are called the __________________ group and the ___________________ group.

47. The __________________ group is subjected to a treatment or condition that the scientist believes is responsible for the observed effect.

48. The control group:

a. must be an exact replicate in every respect
b. must be the same except for the factor being tested
c. is any other group

49. A group of plants is grown on a medium without potassium (a chemical treatment). The plants do poorly. The result of this experiment:

a. proves that potassium is required for plant growth
b. proves that potassium is not required for plant growth
c. proves nothing because there is no control
d. proves nothing because other factors could be responsible for the poor growth
e. Both c and d are correct

50. A concept that provides a logical explanation for a certain body of facts is called a ________________.

51. Theories are:

a. tested in much the same way as are hypotheses (True, False)
b. subjected to if.... then reasoning (True, False)
c. used to predict outcomes (True, False)
d. subject to change (True, False)

52. When we observe that certain events always occur in a precisely predictable way (e.g., an unsupported object always falls) we define such behavior as a ___________________ or natural ___________________.

53. Principles and natural laws are:

a. the same as hypotheses
b. the same as theories
c. descriptions of the way certain things are always observed to behave
d. laws passed by a council of scientists

54. Give two examples of natural laws.

a. ______________________________ b. ______________________________

55. List three uses of instruments in conducting the process of science.

a. ______________________________

b. ______________________________

c. ______________________________

56. Scientists do not use instruments to:

a. increase their power of observation
b. control conditions in various experiments
c. quantify observations
d. avoid making observations

57. Indicate whether the following statements are true [+] or false [-] concerning the process of science.

[ ] There are no controversies or arguments among scientists.
[ ] Progress in science can be slow.
[ ] We are continually confronted by new observations.
[ ] Some observed phenomena may not lend themselves to simple experiments.
[ ] Science is incapable of providing absolute proof for any theory.
[ ] The process of science can be used to test value judgments.
[ ] The validity of science is based on the ability to do experiments.

58, Describe the scientific community.

____

____

59. Give three examples of junk science.

a. ______________________________

b. ______________________________

c. ______________________________

VOCABULARY DRILL

Directions: Match each term with the list of definitions and examples given in the tables below. Term definitions and examples might be used more than once.

Term	Definition	Example	Term	Definition	Example
biodiversity			stewardship		
environmentalist			sustainable		
environmental movement			sustainable development		
environmental racism			sustainable society		
habitat alteration			sustainable yields		

VOCABULARY DEFINITIONS

a. anyone concerned with reducing pollution and protecting wildlife

b. placement of waste sites and other hazardous facilities in towns and neighborhoods where the majority of residents are not white

c. habitat changes that may or may not be reversible

d. the total assemblage of all plants, animals, microbes that inhabit the Earth

e. militant citizenry demanding curtailment and cleanup of pollution and protection of pristine environments

f. continued indefinitely without depleting any of the material or energy resources to keep it running

g. harvest of renewable resources that does not exceed replacement rates

h. continues without depleting its resource base or producing pollutants

i. progress that meets the needs of the present without compromising the needs of future generations

j. an ethic that provides a guide to actions taken to benefit the natural world and other people

VOCABULARY EXAMPLES

a. Rachel Carson	f. urbanization
b. Earth Day 1970	g. American Indians
c. clear-cut logging	h. Our Common Future
d. global flora and fauna	i. saving the spotted owl
e. suburbs of Chattanooga, TN	j. cashing in on the interest only

SELF-TEST

1. Rank (1 = first event to 5 = last event) the following chain of events in the order they would occur in leading to the collapse of a civilization.

Events	Rank
a. loss of topsoil	[ ]
b. population increase beyond food growing capacity	[ ]
c. deforestation	[ ]
d. soil erosion	[ ]
e. disparity between haves and have nots	[ ]

2. Explain why the Easter Island population could have been more susceptible to overpopulation and resource depletion than continental populations?

____

____

3. During which of the following time periods did Americans have a positive attitude toward conservation?

a. scientific era after World War I
b. depression years
c. during World War II
d. after World War II until the 1960s

4. In general, a positive conservation attitude by the majority of Americans is initially stimulated by:

a. public education
b. environmental crisis
c. love of nature
d. federal laws

5. Which of the following was not a post-World War II environmental problem in America?

a. air pollution
b. decline of wildlife populations
c. water pollution
d. overpopulation

6. Indicate whether the following are gains [+] or losses [-] in the war for the environment.

a. [ ] population growth
b. [ ] pollution control
c. [ ] environmental laws
d. [ ] biodiversity

7. Check the most important initial battle to win in the war for the environment.

a. [ ] population growth
b. [ ] pollution control
c. [ ] environmental laws
d. [ ] biodiversity

8. During the spotted owl controversy in the Northwest, the debate was over the _______________________________ versus _______________________________.

9. Match the evidence with list of global events indicating that environmentalism is losing the war to save our planet.

Evidence	Events
[ ] habitat conversions	1. population growth
[ ] CO_2 emissions	2. atmospheric changes
[ ] desertification	3. soil degradation
[ ] 10 billion	4. loss of biodiversity
[ ] commercial exploitation	
[ ] urban development	

10. Any movement toward sustainable development will require public:

a. understanding and awareness of the problems
b. adoption of new values
c. demands for change
d. all of the above

11. Below is a list of worldview assumptions. Check all those that apply to anyone who can be called an environmentalist.

[ ] Natural resources are to be exploited for the advantage of humans.
[ ] Natural resources are essentially infinite.
[ ] We need to stay on the same course of human uses of natural resources as in the past.
[ ] It is alright to spend both the interest and the capital of our natural resource inheritance.
[ ] Natural resources are to be protected and maintained.
[ ] Natural resources are limited by the regenerative capacities of the natural environment.
[ ] We need to change the ways humans use natural resources.
[ ] We need to live only on the interest of our natural resource inheritance.

12. Any actions that promote a sustainable human use of natural resources will require public:

a. understanding and awareness of the problems
b. adoption of new values
c. demands for change
d. all of the above

True [+] or False [-]

13. [ ] American's conservation attitudes have been consistent throughout the twentieth century.
14. [ ] Environmentalists pretty much dictate the future use of global resources.
15. [ ] Air and water in America is cleaner now than in the 1960s.
16. [ ] The outcomes of global trends that are not sustainable are predictable.
17. [ ] The outcomes of establishing a sustainable society are predictable.
18. [ ] The Wise Use and Environmental Movements have compatible goals.
19. [ ] The present American lifestyle is compatible with a sustainable society.
20. [ ] The majority of Americans support the views of contemporary environmentalists.

21. The scientific method begins with:

a. observations b. developing hypotheses c. experimentation d. formulation of theories

22. Hypotheses are confirmed through:

a. observations b. experiments c. authority figures d. technology

23. Experiments must include:

a. experimental groups b. control groups c. controlled environments d. all of these

24. The purpose of the control group is to:

a. show that conditions can be controlled
b. show that factors other than the test factors are not responsible for the observed effect
c. prevent contamination of the experimental group
d. verify the beliefs of authority figures

25. Theories are:

a. tested in much the same way as hypotheses
b. subjected to if . . . . then reasoning
c. subject to change
d. all of the above

26. Events that always occur in a precisely predictable way are called:

a. observations b. hypotheses c. theories d. laws

27. Scientists do not use instruments to:

a. increase their power of observation
b. control conditions in various experiments
c. quantify observations
d. avoid making observations

28. Which of the following is not an accurate statement concerning the process of science?

a. The process of science may lead to answers and more questions.
b. The validity of science is based on the ability to do experiments.
c. There are no controversies or arguments among scientists.
d. Progress in science can be slow.

29. Which of the following is not an accurate statement concerning science?
a. Science can test value judgments.
b. Science can test one-time events.
c. Science can give us answers to all questions.
d. All of the above are not accurate.

ANSWERS TO STUDY GUIDE QUESTIONS

1. closely tied to the land because their survival depends on it; nearly totally isolated from natural world. 2. society fails to care for the environment and sustain it, population increases beyond food growing capacity, disparity between haves and have nots widens; 3. understand how the natural world works, understand how human systems interact with natural systems, assess the status and trends of crucial natural systems, promote and follow a long-term, sustainable relationship with the natural world; 4. population growth and increasing consumption per person, a decline of vital life-support ecosystems, global atmospheric changes, loss of biodiversity; 5. 6 billion, 10 billion; 6. American's consumption of the world's energy and natural resources to produce more possessions is having negative global environmental effects; 7. depleted groundwater supplies, agricultural soils degraded, oceans overfished, forests cut faster than they can grow; 8. global warming; 9. 280 to >375; 10. CO_2 traps heat escaping from the Earth raising the temperature of the Earth's atmosphere; 11. habitat conversion, pollution; 12. habitat alteration, exploitation, pollution; 13. genetic derivatives of all agricultural plants and animals, derivatives of medicinal drugs and chemicals, maintain the stability of natural ecosystems; 14. sustainability, stewardship, sound science; 15. is not; 16. 8K; 17. The 8K represents the surplus - leave the 2K alone to produce another surplus; 18. continues generation after generation, does not exceed sustainable yields, does not pollute beyond nature's capacity to absorb; 19. concerned with growth, efficiency, and maximum use; focuses on human needs and concepts like equity; concerned for preserving integrity of ecosystems, maintaining carrying capacity and dealing with pollution; 20. environmental, social, economic, political; 21. recognize that a trust has been given, responsible care for something not owned, desire to pass something on to future generations; 22. environmental racism; 23. placement of waste site and hazardous facilities in nonwhite neighborhoods; 24. false; 25. wealthy elite horde and squander the resources of developing countries, unjust economic practices of wealthy industrialized countries, patterns of international trade, international debt; 26. false; 27. +, -, +, -, -, +, -; 28. murky and irritating, fowled with raw sewage, decline or extinction; 29. pesticide (DDT) use and high bird mortality; 30. wildlife are the indicators of environmental conditions; 31. Environmental Defense Fund, Greenpeace, Natural Resource Defense Council, Zero Population Growth; 32. Environmental Protection Agency, environmental laws, species saved from extinction, pollution abatement; 33. environment vs. jobs; 34. false; 35. +, -, +, +, +, +; 36. observations; 37. scientific; 38. we perceive reality with our five basic senses; objective reality functions according to certain basic principles and laws; explainable cause and effect; we have the tools and capabilities to understand the basic principles and natural laws; 39. a; 40. sight, smell, taste, sound, touch; 41. rejected; 42. -, +; 43. a; 44. d; 45. 2; 46. experimental, control; 47. experimental; 48. b; 49. e; 50. theory; 51. all true; 52. principle, law; 53. c; 54. Law of Gravity, Conservation of Matter; 55. increase power of observation, control experimental conditions, quantify observations; 56. d; 57. -, +, +, +, +, -, +; 58. A collective body of scientists, working in a given field who, because of their competence and experience, establish what is sound science and what is not; 59. presentation of selective results, political distortions of scientific works, publications in quasi-scientific journals

---

ANSWERS TO SELF-TEST

1. 4, 1, 2, 3, 5 (may be debate on where the chain starts); 2. Population had nowhere else to go, had no import capabilities, landmass small and isolated; 3. b; 4. b; 5. d; 6. -, +, +, -; 7. population growth; 8. environment vs. jobs; 9. 4, 2, 3, 1, 4, 3; 10. d; 11. Checks of the last four and no others are the characteristics of an environmentalist. If you checked a blend of the first and second group of four, you are confused! 12. d; 13. -; 14. -; 15. +; 16. +; 17.+; 18. -; 19. 9. -; 20. + (depending on your opinion of the American public); 21. a; 22. b; 23. d; 24. b; 25. d; 26. d; 27. d; 28. c; 29. d

VOCABULARY DRILL ANSWERS					
Term	**Definition**	**Example**	**Term**	**Definition**	**Example**
biodiversity	d	d	stewardship	j	i
environmentalist	a	a	sustainable	f	f
environmental movement	e	b	sustainable development	i	h
environmental racism	b	e	sustainable society	h	g
habitat alteration	c	c, f	sustainable yields	g	j

CHAPTER 2

ECOSYSTEMS: UNITS OF STABILITY

What Are Ecosystems?

1. All biotic communities consist of ____________________, ____________________, and ____________________ communities.

2. Identify the following community components as abiotic [A] or biotic [B].

   a. [ ] plants   b. [ ] animals   c. [ ] temperature

   d. [ ] microbes   e. [ ] rainfall   f. [ ] wind

3. An organism can be designated a distinct species if it can ____________________ with others of its own kind and produce ____________________ offspring.

4. A group of bullfrogs (<u>Rana catesbiana</u>) in the same farm pond constitutes a ____________________.

5. An ecosystem may be defined as a grouping of ____________________, ____________________, and ____________________ interacting with ____________________ and their ____________________.

6. Furthermore, the interrelationships are such that the grouping may be ____________________.

7. List six examples of major kinds of ecosystems.

   a. ____________________ b. ____________________ c. ____________________

   d. ____________________ e. ____________________ f. ____________________

8. Each ecosystem is characterized by distinctive plant ____________________.

9. Identify the ecotone in Fig. 2-2. ____________________

10. The major ecosystem of the eastern United States is ____________________.

11. The major ecosystem of the central United States is ____________________.

12. The major ecosystem of the southwestern United States is ____________________.

13. The distribution of major ecosystems globally is the consequence of ____________________.

1[illegible] [illegible] in one ecosystem or biome may affect other ecosystems or biomes.

15. Give two examples of freshwater aquatic ecosystems.

   a. ____

   b. ____

16. What type of aquatic ecosystem requires a mixture of fresh and salt water?
____________________.

The Structure of Ecosystems

17. The two basic structural components of all ecosystems are the ____________________ communities and ____________________ environmental factors.

Trophic Categories

18. All the organisms in an ecosystem can be assigned to one of three categories. These three categories are ____________________, ____________________, and ____________________.

19. Producers include all plants that do what? ____________________

20. In photosynthesis, plants use ________________ energy to convert ________________ gas and ________________ into ________________.

21. The word inorganic refers to substances such as ________________.

22. The chemicals that make up the bodies of organisms are referred to as ________________.

23. In short, producers convert (inorganic, organic) chemicals into (inorganic, organic) chemicals.

24. The energy for this production comes from ________________.

25. Organisms that can produce all their own organic materials from inorganic raw materials in the environment are called (autotrophs, heterotrophs). Give an example. ________________

26. Organisms that must consume organic material as a source of nutrients and energy are called (autotrophs, heterotrophs). Give an example. ________________

27. (True or False) All plants are autotrophs. If you said false, give an example of an exception.

____

28. List three organisms that could be classified as consumers.

a. ________________ b. ________________ c. ________________

29. Consumers that feed directly on producers are called primary consumers or ________________.

30. List two organisms that are primary consumers.

a. ________________ b. ________________

31. Consumers that feed on other consumers are called secondary consumers or ________________ or ________________.

32. Indicate whether the following list of animals would be predators [PR] or prey [PY] in predator/prey relationships.

a. [ ] wolf b. [ ] coyote c. [ ] quail
d. [ ] bear e. [ ] deer f. [ ] salmon

33. Give an example of:

a. internal parasite: ________________ b. external parasite: ________________

c. parasite host: ________________

34. All dead organic matter (dead leaves, twigs, bodies of animals, fecal matter) is called ________________.

35. Give three examples of detritus feeders.

a. ____________________ b. ____________________ c. ____________________

36. Decomposers consist of __________________ and ____________________.

Trophic Relationships: Food Chains, Food Webs, and Trophic Levels

37. Construct a food chain using a grassland community that consists of [a] big bluestem grass, [b] grasshoppers, [c] frogs, [d] snakes, and [e] owls. First consider what each organism eats. For example:

a. grasshoppers eat: __________________ b. frogs eat: __________________

c. snakes eat: __________________ d. owls eat: __________________

e. Next, place each animal in a sequence of feeding relationships [a to e] that demonstrates a food chain.

[ ] , [ ] , [ ] , [ ] , [ ]

38. Let's add some more organisms to the above grassland community, including fescue grass, berry bushes, mice, white-tailed deer, humans, rabbits, foxes, red-tailed hawk, ground squirrels, and sunflowers. Construct a food web in the diagram below by placing each species in its proper trophic level and then drawing lines of feeding relationships (see Fig. 2-13 for an example of this exercise). The species to use in this exercise are as follows: a. big bluestem grass, b. grasshoppers, c. frogs, d. snakes, e. owls, f. fescue grass, g. berry bushes, h. mice, i. deer, j. humans, k. rabbits, l. foxes, m. hawks, n. ground squirrels, and o. sunflowers.

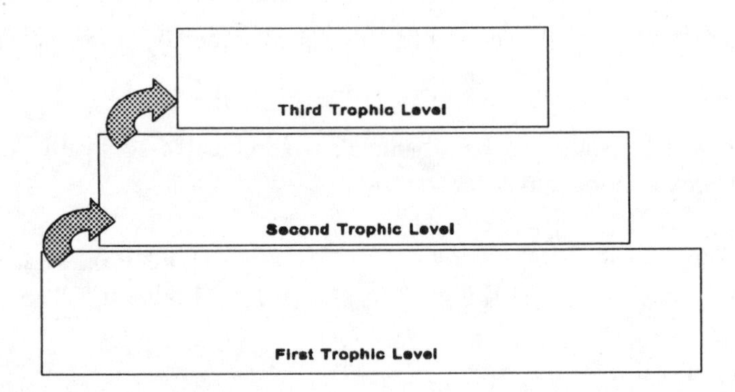

39. Indicate whether the following categories of organisms occupy the [a] first, [b] second, [c] third, or [d] more than one trophic level.

a. [ ] producers b. [ ] consumers c. [ ] decomposers

d. [ ] carnivores e. [ ] herbivores f. [ ] predators

g. [ ] prey h. [ ] omnivores i. [ ] parasites

j. [ ] autotrophs k. [ ] heterotrophs

40. How is biomass determined?

____

____

41. In an ecosystem the biomass of herbivores is always (greater, less) than the biomass of producers, and the biomass of carnivores is always (greater, less) than the biomass of herbivores.

42. Diagram the biomass relationship for three trophic levels.

Nonfeeding Relationships

43. Describe an example of mutualism.

____

44. How does your example of mutualism demonstrate the concept of symbiosis?

____

45. Where an organism lives is called its (habitat or niche). How an organism lives is called its (habitat or niche).

46. Give an example of how interspecies niche competition is reduced in:

a. space: ____

b. time: ____

Abiotic Factors

47. Three examples of abiotic factors are:

a. ______________ b. ______________ c. ______________

48. Label the figure below with the following terms (see Fig. 2-17).

a. optimum b. range of tolerance c. limits of tolerance d. zones of stress

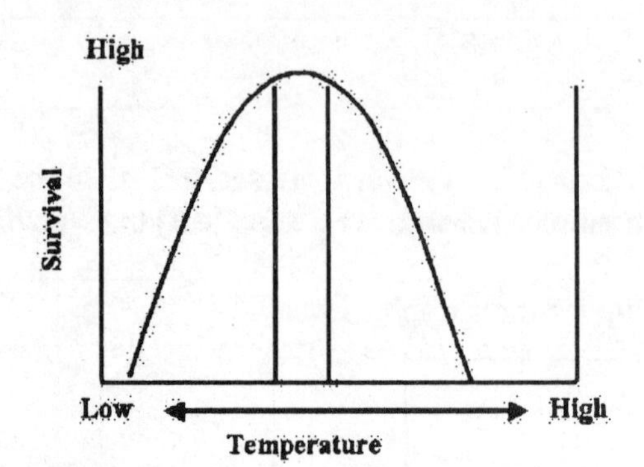

49. As the amount or level of any factor increases or decreases from the optimum, the plant or animal is increasingly ______________________.

50. If any factor increases or decreases beyond an organism's limit of tolerance, what occurs?

______________________.

51. (True or False) Each species has an optimum zone of stress and limits of tolerance with respect to every factor.

52. How many limiting factors need to be beyond a plant's limit of tolerance to preclude its growth?

____

53. Give two examples of limiting factors.

a. ______________________ b. ______________________

Global Biomes

The Role of Climate

54. The definition of climate involves both ____________________ and____________________.

55. ____________________ is the main climatic factor responsible for the separation of ecosystems into forests, grasslands, and deserts.

56. Even though it takes 30 inches or more of rainfall to support a forest ecosystem, __________________ will determine the kind of forest.

57. Give an example of an ecosystem that would contain permafrost. ____________________

Microclimate and Other Abiotic Factors

58. A localized area within an ecosystem that has moisture and temperature conditions different from the overall climate of the area is called a __________________.

Biotic Factors

59. The biotic factor that limits grass from taking over high rainfall regions is

a. tall trees b. herbivores c. parasites d. insects

Physical Barriers

60. The spread of species into other ecosystems may be prevented by natural physical barriers, such as ______________, ______________, and ______________.

61. The spread of species into other ecosystems may be prevented by physical barriers produced by humans, such as ______________, ______________, and ______________.

Implications for Humans

Three Revolutions

62. As hunter-gatherers, humans were much like: a. herbivores b. carnivores c. omnivores

63. With ______________, humans created their own distinctive ecosystem apart from natural ecosystems.

64. List five ways humans have overcome the usual limiting factors that prevent the spread of our species.

a. ______________________________

b. ______________________________

c. ______________________________

d. ______________________________

e. ______________________________

65. (True or False) The human ecosystem is upsetting and destroying other ecosystems.

VOCABULARY DRILL

Directions: Match each term with the list of definitions and examples given in the tables below. Term definitions and examples may be used more than once.

Term	Definition	Example(s)	Term	Definition	Example(s)
abiotic			heterotroph		
abiotic factors			host		
autotroph			inorganic molecule		
biomass			limiting factors		
biomass pyramid			limits of tolerance		
biome			microclimate		
biosphere			mutualism		
biota			niche		
biotic community			omnivore		
biotic factors			optimum		
biotic structure			organic molecule		
carnivore			parasites		
chlorophyll			parasitism		
climate			photosynthesis		
consumers			population		
decomposer			predator		
detritus			prey		
detritus feeders			primary consumer		
ecologist			producers		
ecology			range of tolerance		
ecosystem			secondary consumer		
ecotone			species		
food chain			symbiotic		
food web			synergistic effect		
habitat			trophic levels		
herbivore			zone of stress		

VOCABULARY DEFINITIONS
a. grouping or assemblage of living organisms in an ecosystem
b. assemblage of nonliving physical or chemical components in an ecosystem
c. specific kinds of plants, animals, or microbes that interbreed and produce fertile offspring
d. number of individuals that make up an interbreeding, reproducing group within a given area
e. a grouping of plants, animals, and microbes interacting with each other and their environment in such a way to perpetuate the grouping
f. study of ecosystems and interactions
g. people who study ecosystems and interactions
h. the transitional area where one ecosystem blends into another
i. total combined dry weight of all organisms at a trophic level
j. all the species and ecosystems combined
k. the way different categories of organisms fit together in an ecosystem
l. organisms that produce their own organic material from inorganic constituents
m. organisms that must feed on complex organic material to obtain energy and nutrients
n. the process of converting sunlight energy and CO_2 to sugar and O_2
o. molecule that plants use to capture light energy for photosynthesis
p. constructed in large part from carbon and hydrogen atoms
q. constructed in large part from elements other than carbon and hydrogen
r. dead or partially digested plant or animal material
s. primary detritus feeders
t. feed directly on producers
u. feed on primary consumers
v. meat-eating secondary consumer
w. plant and meat-eating consumer
x. animal that attacks, kills, and feeds on another animal
y. animal killed and eaten by a predator
z. a predator that feeds off its prey for a long time, typically without killing it
aa. plant or animal fed upon by a parasite
bb. pathways where one organism is eaten by a second, which is eaten by a third, and so on
cc. a way of demonstrating the interrelatedness of food chains
dd. feeding levels within a food chain or web

VOCABULARY DEFINITIONS
ee. shape of biomass potential at each trophic level
ff. a symbiotic relationship between two organisms in which both derive benefit
gg. living together
hh. a symbiotic relationship between organisms in which one benefits and the other is harmed
ii. plant community and physical environment where an organism lives
jj. what an organism feeds on, where and when it feeds, where it finds shelter and nesting sites
kk. physical and chemical components of ecosystems
ll. a certain abiotic level where organisms survive best
mm. abiotic level between optimal range and high and low level of tolerance
nn. high and low ranges of tolerance to abiotic levels
oo. any factor that limits the growth, reproduction, and survival of organisms
pp. greater effect of two factors interacting together than individually
qq. description of average temperature and precipitation each day throughout the year
rr. conditions in specific localized areas
ss. limiting factors caused by other species
tt. grouping of related ecosystems into major kinds of ecosystems
uu. the entire span of abiotic values that allow any growth at all
vv. organisms that feed primarily on dead, decaying, or partially digested organic matter

VOCABULARY EXAMPLES	
a. plants, animals, microbes	q. salt (NaCl)
b. CO_2, rainfall, pH, O_2, salinity, temperature	r. dead leaves
c. Rana catesbiana (bullfrogs)	s. bacteria or fungi
d. All Rana catesbiana in a farm pond	t. deer or rabbit
e. Fig. 2-2	u. wolf or owl
f. what you are doing now	v. pigs and bears
g. your teacher?	w. Fig. 2-7
h. where grassland meet forest	x. Fig. 2-11a
i. Fig. 2-12	y. Figs. 2-14 and 2-15
j. Earth	z. Fig. 2-8 or Fig. 2-9
k. producers, consumers, decomposers	aa. Fig. 2-17
l. green plants	bb. chemical A → effect B, chemical C → effect D, chemical A + B → effect E
m. herbivores, carnivores, omnivores	cc. Fig. 2-20
n. $6CO_2 + 6H_2O \rightarrow C_6H_{12}O_6 + 6O_2$	dd. Fig. 2-21
o. green organelle in plant tissue	ee. Table 2 -3
p. glucose ($C_6H_{12}O_6$)	

SELF - TEST

An ecosystem is best defined as a grouping of 1.________________, 2.________________ and 3. ________________ interacting with 4. ________________ and their 5. ________________ in such a way that the grouping is perpetuated.

Ecosystems evolve in sequential patterns involving plant, animal, and abiotic components. What is the component sequence in ecosystem development?

6. First Component: ________________ 7. Second Component: ________________

8. Third Component: ________________

9. A major biome in the central United States is a :

   a. deciduous forest b. grassland c. desert d. coniferous forest

10. Which of the following is not correctly matched?

    a. oak tree = producer
    b. squirrel = consumer
    c. mushroom = detritus feeder
    d. fungi = decomposers

11. The three major biotic components of ecosystem structure are

    a. producers, herbivores, and carnivores
    b. producers, consumers, and decomposers
    c. plants, animals, and climate
    d. consumers, detritus feeders, and decomposers

Below is a typical food chain illustration. Answer the series of questions below the illustration by choosing level 1, 2, 3, 4, 5, or ALL.

Level 5: Owl
Level 4: Snake
Level 3: Frog
Level 2: Grasshopper
Level 1: Grass

At which level:

12. _____ is the most transferable energy available?

13. _____ would one find the least biomass?

14. _____ is a primary consumer located?

15. _____ is a producer located?

16. _____ could a white-tailed deer be a representative?

17. _____ is sunlight the original energy source?

18. _____ would one find decomposers?

19. _____ would one find parasites?

20. The biomass relationship of a food chain can only be depicted as a: a. box, b. rectangle, c. circle, d. pyramid

21. Explain your answer choice to question 20.

____

22. Which of the following is not an example of an abiotic factor?

a. water b. light c. dead log d. temperature

23. The points at which a factor becomes so high or low as to threaten an organism's survival is referred to as

a. range of optimums b. range of limits c. range of tolerance d. limits of tolerance

24. How many factors need to be beyond an organism's limits of tolerance to cause stress?

a. one b. five c. ten d. hundreds

Which limiting factor (a. temperature, b. rainfall, c. pH, d. soil type) is the best determinant of:

25. _____ deserts, grasslands, and forests?

26. _____ kind of forest?

27. _____ kind of desert?

Identify those conditions that do [+] and do not [-] represent the human system and then indicate whether the listed conditions are [+] or are not [-] compatible with natural ecosystems.

Human System Conditions	Compatibility with Natural Ecosystems
28. [ ] Recycle wastes	32. [ ]
29. [ ] Practice population control	33. [ ]
30. [ ] Control limiting factors	34. [ ]
31. [ ] Cause rapid abiotic/biotic changes	35. [ ]

ANSWERS TO STUDY GUIDE QUESTIONS

1. plants, animals, microbe; 2. B, B, A, B, A, A; 3. reproduce, viable; 4. population; 5. plants, animals, microbes, each other, environment; 6. perpetuated; 7. deciduous forest, grassland, deserts, coniferous forest, tundra, tropical rain forest; 8. communities; 9. grassland/forest; 10. deciduous forest; 11. grassland; 12. desert; 13. climate (temperature and annual rainfall); 14. true; 15. lakes and ponds, streams and rivers, inland wetlands; 16. estuaries; 17. biotic, abiotic; 18. producers, consumers, decomposers/detritus feeders; 19. carry on photosynthesis; 20. light, carbon dioxide, water, glucose; 21. air, rocks, or water; 22. organic; 23. inorganic, organic; 24. sunlight; 25. autotrophs, green plants; 26. heterotrophs, animals and fungi; 27. false, Indian pipe; 28. humans, fish, worms; 29. herbivores; 30. deer, mice; 31. carnivores, omnivores; 32. PR, PR, PY, PR, PY, PY; 33. tapeworm, tick, any living organism; 34. detritus; 35. earthworms, clams, crayfish; 36. fungi, bacteria; 37. grass, grasshoppers, frogs, snakes, a, b, c, d, e; 38. First trophic level: a, f, g, o, Second trophic level: b, h, i, k, n, Third trophic level: c, d, e, j, l, m; Example of food web relationship would be to link all the Second trophic level organisms with all the First trophic level organisms; 39. a, d, d, c, b, c, b, d, d, a, d; 40. total dry weight of all organisms occupying a trophic level; 41. less, less; 42. illustrations should be a pyramid; 43. relationship between bees and flowers; 44. bees and flowers require a close union to accomplish life functions; 45. habitat, niche; 46. birds feeding in different parts of a tree, day versus night time feeders on same food source; 47. wind, temperature, precipitation; 48. see Fig. 2-17; 49. stressed; 50. death; 51. true; 52. one; 53. light, temperature, water; 54. temperature, precipitation; 55. annual precipitation; 56. temperature; 57. alpine tundra; 58. microclimate; 59. a; 60. ocean, desert, mountain range; 61. roadways, dams, cities; 62. c; 63. agriculture; 64. producing abundant food, creating reservoirs and distributing water, overcoming predation and disease, constructing our own habitats, overcoming competition with other species; 65. true

---

ANSWERS TO SELF-TEST

1. plants; 2. animals; 3. microbes; 4. each other; 5. environment; 6. abiotics; 7. plants; 8. animals; 9. b; 10. c; 11. b; 12. 1; 13. 5; 14. 2; 15. 1; 16. 2; 17. 1; 18. all; 19. all; 20. d; 21. energy loss from one level to the next decreases potential for biomass production at higher levels; 22. c; 23. d; 24. a; 25. b; 26. a; 27. a; 28. -; 29. -; 30. +; 31. +; 32. +; 33. + 34. -; 35. -

VOCABULARY DRILL ANSWERS					
Term	**Definition**	**Example(s)**	**Term**	**Definition**	**Example(s)**
abiotic	b	b	heterotroph	m	c,m,s,t,u,v,w
abiotic factors	kk	b	host	aa	a,c,g,k,l,m,s,t, u,v,w
autotroph	l	l	inorganic molecule	q	q
biomass	i	i	limiting factors	oo	b
biomass pyramid	ee	x	limits of tolerance	uu	aa
biome	tt	ee	microclimate	rr	dd
biosphere	j	j	mutualism	ff	y
biota	a	a	niche	jj	z
biotic community	a	a	omnivore	w	v
biotic factors	ss	a	optimum	ll	aa
biotic structure	k	k	organic molecule	p	p
carnivore	v	u	parasites	z	w
chlorophyll	o	o	parasitism	hh	w
climate	qq	cc	photosynthesis	n	n
consumers	m	m	population	d	d
decomposer	s	s	predator	x	u
detritus	r	r	prey	y	c, t
detritus feeders	vv	s	primary consumer	t	t
ecologist	g	g	producers	l	l
ecology	f	f	range of tolerance	uu	aa
ecosystem	e	e	secondary consumer	u	u
ecotone	h	e, h	species	c	c
food chain	bb	x	symbiotic	gg	y
food web	cc	z	synergistic effect	pp	bb
habitat	ii	e	trophic levels	dd	x
herbivore	t	t	zone of stress	mm	aa

CHAPTER 3

ECOSYSTEMS: HOW THEY WORK

Matter, Energy and Life

1. Matter, in a chemical sense, refers to all ____________________, ____________________, and ____________________.

2. Atoms or elements are the basic building blocks of all ____________________.

3. How many different kinds of elements occur naturally? ____________________

4. Only __________ of all the known elements are used to produce living tissues. (See Table 3-2; iodine is missing)

5. Indicate whether the following statements are true [+] or false [-] concerning the Law of Conservation of Matter.

[ ] Atoms cannot be created.
[ ] Atoms can be destroyed.
[ ] Atoms can be converted into others.
[ ] Atoms can be rearranged to form different molecules.

Matter in Living and Nonliving Systems

6. (True or False) A molecule = a compound. (True or False) A compound = a molecule.

7. Indicate whether each of the following is an element [E] or compound [C].

a. [ ] oxygen b. [ ] carbon dioxide c. [ ] water d. [ ] nitrogen
e. [ ] ozone f. [ ] phosphorus g. [ ] carbon h. [ ] protein
i. [ ] sugar j. [ ] hydrogen

8. For each of the following essential elements in living organisms, give a compound or molecule that contains it and where it is found in the abiotic environment (see Table 3-1).

Element	Compound Containing It	Environmental Location
carbon	____________	____________
hydrogen	____________	____________
oxygen	____________	____________
nitrogen	____________	____________
phosphorus	____________	____________

9. Match the list of elements and molecules.

Elements	Molecules
a. N ________	1. glucose
b. C ________	2. proteins
c. H ________	3. starch
d. O ________	4. fats
e. P ________	5. nucleic acids
f. S ________	6. all of the above

10. (True or False) All organic compounds are found in living systems.

Energy Considerations

11. Matter is defined as anything that occupies ________________ and has ________________.

12. From the most to least dense states, matter can exist in the forms of ____________________, ____________________ and ____________________.

13. **Energy** is defined as: ____

14. List three forms of energy: ____________________, ____________________, and ____________________.

15. Indicate whether the following statements about energy are true [+] or false [-].

[ ] Energy occupies space.
[ ] Energy can be weighed.
[ ] Kinetic energy can be transformed into potential energy.
[ ] Potential energy can be transformed into kinetic energy.
[ ] Energy can be recycled.
[ ] Energy can be created.
[ ] Energy can be converted from one form to another.
[ ] There is some loss in each energy conversion.
[ ] Energy can be measured.

16. Heat, light, and motion are forms of (kinetic, potential) energy.

17. Glucose, a compressed spring, and fossil fuels are forms of (kinetic, potential) energy.

18. The high potential energy in fuels is commonly referred to as ____________________ energy.

19. A ____________________ is the unit measure of the amount of heat required to raise the temperature of 1 gram of water 1 degree Celsius.

20. Indicate whether the following statements pertain to the [a] First or [b] Second Law of Thermodynamics.

[ ] Energy cannot be created or destroyed.
[ ] Energy can be converted from one form to another.
[ ] Energy conversions always result in net losses of useful energy.
[ ] Heat is the final energy conversion form.

Energy Changes in Organisms and Ecosystems

21. Going from left to right, identify whether the level of potential energy in the following types of molecules or matter is high [H] or low [L].

a. [ ] inorganic b. [ ] organic c. [ ] coal
d. [ ] water e. [ ] sugar f. [ ] wood

22. In most ecosystems, the primary producers are plants that carry on ____________________.

23. Light energy for photosynthesis is absorbed by the ____________________ molecule.

24. The efficiency of photosynthesis in converting light energy to chemical energy is at best __________ percent efficient.

25. The primary product of photosynthesis is ________________.

26. List three ways plants use the glucose produced during photosynthesis.

a. ____

b. ____

c. ____

27. The basic source of energy and raw materials for animal growth and development is

____

28. Cell respiration is effectively the ________________ process of photosynthesis in terms of what is consumed and released.

29. Indicate whether each of the following refers to photosynthesis [P], cell respiration [R], both [B], or neither [N].

[ ] releases oxygen
[ ] stores energy
[ ] releases carbon dioxide
[ ] consumes carbon dioxide
[ ] releases energy
[ ] produces sugar
[ ] consumes sugar
[ ] consumes oxygen

30. Place an [x] by any of the following organisms that carry on cell respiration for at least part of their energy needs.

[ ] consumers [ ] decomposers [ ] plants [ ] fungi [ ] bacteria

31. The lack of certain nutrients in your diet may lead to various diseases caused by

________________.

32. The point of a balanced diet is that it supplies both ________________ and ________________ in adequate but not excessive amounts.

33. Trace the three pathways of food eaten by any consumer.

a. ____
b. ____

c. ____

34. (True or False) Fecal waste is no longer a source of nutrients and energy.

35. Give two examples of detritus feeders that require a mutualistic relationship with other organisms to digest cellulose.

a. ______________________ b. ______________________

Principles of Ecosystem Function

Energy Flow in Ecosystems

36. What three factors account for energy loss from one trophic level to another?

a. ____

b. ____

c. ____

37. There is an estimated _____ percent loss of energy as it moves from one trophic level to the next.

38. Biomass (increases, decreases) at each successive trophic level.

39. The energy available to producers is (greater, less) than that available to herbivores; therefore, the biomass of producers will be (greater, less) than that of herbivores.

40. The energy available to herbivores is (greater, less) than that available to carnivores; therefore, the biomass of herbivores will be (greater, less) than that of carnivores.

41. Indicate whether the following terms pertain to the first [1], second [2], third [3], fourth [4], or fifth [5] principle of ecosystem sustainability (see Table 3-1).

a. [ ] biomass
b. [ ] recycling
c. [ ] solar energy
d. [ ] nonpolluting
e. [ ] carbon cycle
f. [ ] food chains
g. [ ] restoration ecology

42. Explain why the illustration of total biomass at different trophic levels is a pyramid.

____

____

43. Eventually all of the energy escapes as ______________________ energy.

44. (True or False) Energy can be recycled.

45. For sustainability, ecosystems use sunlight as their source of energy is the (First, Second, Third) Principle of Ecosystem Sustainability.

46. Identify two benefits derived from solar energy.

a. ______________________ b. ______________________

Biogeochemical Cycles

47. For sustainability, ecosystems dispose of wastes and replenish nutrients by recycling all elements is the (First, Second, Third) Principle of Ecosystem Sustainability.

48. Natural ecosystems do not harm themselves with their own waste products because these products are ______________________.

The Carbon Cycle

49. The reservoir of carbon for the carbon cycle is ______________________ present in the ______________________.

50. Carbon dioxide is first incorporated into sugar by the process of ______________________, and then it may be passed to other organic compounds in other organisms through food chains.

51. At any point on the food chain an organism may use the carbon compounds in cell respiration to meet its energy needs. When this occurs, the carbon atoms are released as ______________________.

52. The amount of atmospheric CO_2 is being increased greatly by the burning of ______________________.

53. Complete the following diagram of the Carbon Cycle by putting in all arrows of interrelationships between organisms and labeling the empty boxes. See Fig. 3-16 for an example of this cycle.

The Phosphorus Cycle

54. The reservoir of the element phosphorus exists in various rock and soil ________________.

55. Plants absorb phosphate from the __________________ or ____________________ solution.

56. Phosphate is first assimilated into organic compounds as ____________________ phosphate and then passed to various other organisms through the food chain.

57. At any point on the food chain, an organism may use organic phosphorus in respiration. When this occurs, the phosphate is released back to the environment in ________________.

58. Humans disrupt the phosphorus cycle by _________________________ and _________________________.

59. Complete the following diagram of the Phosphorus Cycle by putting in all arrows of interrelationships between organisms and labeling the empty boxes. See Fig. 3-17 for an example of this cycle.

The Nitrogen Cycle

60. The main reservoir of N is in the air, which is about __________ percent nitrogen gas.

61. (True, False) Plants cannot assimilate nitrogen gas from the atmosphere.

62. To be assimilated by higher plants, nitrogen must be present as ____________________ or ____________________ ions.

63. A number of bacteria and certain blue green algae can convert nitrogen gas to the ammonium form, a process called nitrogen __________________.

64. The bacteria have a mutualistic relationship with the ________________ family of plants.

65. Once fixed, nitrogen may be passed down food chains, excreted as _________________, and recycled as the mineral nutrient called _________________.

66. Nitrogen eventually returns back to the atmosphere because certain bacteria convert the nitrate and ammonia compounds back to ___________________.

67. Complete the following diagram of the Nitrogen Cycle by putting in all arrows of interrelationships between organisms and labeling the empty boxes. See Fig. 3-18 for an example of this cycle.

68. (True or False) All natural terrestrial ecosystems are dependent on the presence of nitrogen-fixing organisms and legumes.

69. The human agricultural system bypasses nitrogen fixation by using fertilizers containing ________________ or ____________________.

Implications for Humans

Sustainability

70. Indicate whether the following human actions are a violation of the first [1] of second [2] principle of ecosystem sustainability.

[ ] lack of recycling
[ ] excessive use of fossil fuels
[ ] feeding largely on the third trophic level
[ ] excessive use of fertilizers
[ ] use of nuclear power
[ ] using agricultural land to produce meat
[ ] destruction of tropical and other rain forests
[ ] nutrient overcharge into marine and freshwater ecosystems
[ ] production and use of nonbiodegradable compounds

Value

71. Natural capital is identified by the ____________________ and ____________________ represented in natural ecosystems.

72. List 5 of the 17 major ecosystem goods and services.

a. ____________________ b. ____________________

c. ____________________ d. ____________________

e. ____________________

73. What is the estimated total value to human welfare for one year of natural ecosystem goods and services? __________ trillion dollars

Managing Ecosystems

74. What is the goal of ecosystem management?

____

75. What does the goal of ecosystem management mean?

____

VOCABULARY DRILL

Directions: Match each term with the list of definitions and examples given in the tables below. Term definitions and examples may be used more than once.

Term	Definition	Example(s)	Term	Definition	Example(s)
anaerobic			malnutrition		
atoms			matter		
cellulose			mineral		
chemical energy			mixture		
compound			molecule		
elements			natural organics		
energy			organic molecule		
entropy			organic phosphate		
fermentation			overgrazing		
First Law of Thermodynamics			oxidation		
inorganic molecule			potential energy		
kinetic energy			Second Law of Thermodynamics		
Law of Conservation of Matter			synthetic organics		

VOCABULARY DEFINITIONS
a. anything that occupies space and has mass
b. basic building blocks of all matter
c. any two or more atoms bound together
d. any two or more different atoms bound together
e. no chemical bonding between molecules involved
f. any hard crystalline material of a given composition
g. carbon-based molecules that make up the tissue of living organisms
h. molecules with neither carbon-carbon nor carbon-hydrogen bonds
i. compounds produced by living organisms
j. compounds produced by humans in test tubes

VOCABULARY DEFINITIONS
k. the ability to move matter
l. energy in action or motion
m. energy in storage
n. potential energy contained in fuels
o. energy is neither created nor destroyed but may be converted from one form to another
p. in any energy conversion, you will end up with less usable energy that you started with
q. disorder
r. systems will go spontaneously in one direction only toward increasing entropy
s. process of oxidizing glucose in the presence of oxygen into carbon dioxide and water
t. process of breaking down molecules
u. absence of one or more nutrients in the diet
v. stored form of glucose in plant cell walls
w. partial oxidation of glucose that can occur in the absence of oxygen
x. oxygen-free environment
y. phosphate bound in organic compounds
z. a form of nonsustainable harvest
aa. atoms are not created or destroyed in chemical reactions

VOCABULARY EXAMPLES		
a. gases, liquids, solids	j. burning match	s. alcohol
b. NCHOPS	k. a potato	t. deep space
c. O_2	l. gasoline	u. nucleotides
d. CH_4	m. You can't get something for nothing.	v. clear-cut logging
e. muddy water	n. You can't even break even.	w. $C_6H_{12}0_6 + 60_2 \rightarrow 6C0_2 + 6H_20$
f. diamond	o. decomposition	
g. NaCl	p. digestion	
h. plastics	q. vitamin deficiency	
i. walking	r. fiber	

SELF-TEST

1. Identify 17 elements out of the 108 (Table C-1) that are essential in the structure or function of living organisms? (Your text gave you 16 in Table 3-2.)

The essential elements of life include:____

____

If a nonessential element becomes incorporated into a living system (e.g., yourself), what are three potential impacts of this element insult on the living system?

2. ____

3. ____

4. ____

Identify from the examples of organization of matter on the right the most simple and the most complex levels.

5. [ ] most simple

6. [ ] most complex

a. organisms
b. proteins, carbohydrates, fats
c. atoms
d. cells

7. What is the carbon source for photosynthesis? ________________

8. What is the energy source for photosynthesis? ________________

9. What is the carbon source for cellular respiration? ________________

10. What is the energy source for cellular respiration? ________________

Use the following chemical equations to answer the next series of questions.

a. $6CO_2 + 6H_2O \rightarrow C_6H_{12}O_6 + 6O_2$

b. $C_6H_{12}O_6 + 6O_2 \rightarrow 6CO_2 + 6H_2O$

Which equation:

11. [ ] is the chemical formula for photosynthesis?

12. [ ] is the chemical formula for cellular respiration?

13. [ ] shows glucose as an end product?

14. [ ] shows the raw materials for photosynthesis as the end product?

15. [ ] uses a kinetic energy source?

16. [ ] uses a potential energy source?

17. How does equation [a] demonstrate the Law of Conservation of Matter?

____

____

Associate the following terms with the [a] First Law or [b] Second Law of Thermodynamics.

18. [ ] entropy

19. [ ] loss

20. [ ] change

21. In every energy conversion:

a. some energy is converted to heat
b. heat energy is lost
c. lost heat may be recaptured and reused
d. both a and b

22. Consumers must feed on preexisting organic material to obtain:

a. nutrients, b. energy, c. both a and b, d. neither a or b; consumers make their own nutrients and energy

23. Which of the following is not recycled in natural ecosystems?

a. nitrogen b. carbon c. energy d. phosphorus

24. The carbon atoms in the food you eat will

a. be exhaled as carbon dioxide
b. become part of your body tissues
c. pass through as fecal wastes
d. a, b, and c

25. In the phosphorus cycle, soil phosphate first enters the ecosystem through:

a. plants b. herbivores c. carnivores d. decomposers

26. Nitrogen fixation refers to:

a. releasing nitrogen into the air
b. converting it to a chemical form plants can use
c. repairing broken molecules
d. applying fertilizer

27. The nitrogen cycle relies on

a. mutualism b. commensalism c. parasitism d. synergism

Indicate whether the carbon [C], nitrogen [N], phosphorus [P], or more than one [M] cycle has been negatively effected by the following human actions.

28. [ ] abandonment of rotation cropping

29. [ ] heavy reliance on fossil fuels

30. [ ] discharges of household detergents into natural waterways

31. Which Law explains why it is impossible to have a greater biomass at the highest level in a food chain?

a. Law of Conservation of Matter,
b. First Law of Thermodynamics, or
c. Second Law of Thermodynamics

32. How does this law (your response to question 31) explain why it is impossible to have a greater biomass at the highest level in a food chain?

____

____

33. Explain why it makes sense to nurture a recycling mentality in the human system.

____

ANSWERS TO STUDY GUIDE QUESTIONS

1. gases, liquids, solids; 2. life; 3. 92; 4. 17; 5. +, +, -, +; 6. true, false; 7. E, C, C, E, C, E, E, C, C, E; 8. (carbon dioxide, water), (water, water), (oxygen gas, air), (nitrogen gas, air), (phosphate ion, dissolved in water or in rock or soil minerals); 9. 2 and 5, 6, 6, 6, 4 and 5, 2 and 5; 10. false; 11. space, mass; 12. solids, liquids, gas; 13. ability to move matter; 14. kinetic, potential, chemical; 15. -, -, +, +, -, -, +, +, +; 16. kinetic; 17. potential; 18. chemical; 19. calorie; 20. a, a, b, b; 21. L, H, H, L, H, H; 22. photosynthesis; 23. chlorophyll; 24. 2-5; 25. glucose; 26. raw materials for production of other organic molecules, plant cell respiration, stored for future use; 27. glucose; 28. reverse; 29. P, B, R, P, B, P, B, B; 30. all are marked; 31. malnutrition; 32. calorie, nutrients; 33. 60 to 90 percent used to produce energy, body growth and maintenance, undigested residue passes out as wastes; 34. false; 35. termites, cows; 36. much of the preceding trophic level is standing biomass and is not consumed, much of what is consumed is used for energy, some of what is consumed is undigested and passes through the organism; 37. 90; 38. decreases; 39. greater, greater; 40. greater, greater; 41. 3, 2, 1, 1, 2, 1, 4; 42. because of energy loss from lower to higher trophic levels; 43. heat; 44. false; 45. first; 46. nonpolluting, nondepleting; 47. second; 48. recycled; 49. carbon dioxide, air; 50. photosynthesis; 51. carbon dioxide; 52. fossil fuels; 53. see Fig. 3-16; 54. minerals; 55. soil, water; 56. organic; 57. urine and other wastes; 58. cutting down rain forests, phosphate fertilizer runoff from agricultural lands; 59. See Fig. 3-17; 60. 78; 61. true; 62. ammonium, nitrates; 63. fixation; 64. legume; 65. ammonium, nitrates; 66. nitrogen gas; 67. See Fig. 3-18; 68. true; 69. ammonium, nitrate; 70. 2, 1, 1, 2, 1, 1, 1, 2, 2; 71. goods, services; 72. See Table 3-3; 73. 33; 74. manage ecosystems to assure their sustainability; 75. maintaining ecosystems such that they continue to provide the same level of goods and services indefinitely into the future

---

ANSWERS TO SELF-TEST

1. B, Ca, C, Cl, Cu, H, I, Fe, Mg, Mn, Mo, N, O, P, K, Na, S, Zn; 2. death; 3. the element might be stored in body fat or organs; 4. the element may be discharged with other wastes; 5. c; 6. a; 7. carbon dioxide; 8. sunlight; 9. glucose; 10. glucose; 11. a; 12. b; 13. a; 14. b; 15. a; 16. b; 17. the same number of C, H, and O atoms that go into the reaction come out at the end of the reaction; 18. b; 19. b; 20. a; 21. d; 22. c; 23. c; 24. d; 25. a; 26. b; 27. a; 28. N; 29. C; 30. P; 31. c; 32. energy loss from one level to the next decreases potential for biomass production at higher levels; 33. because all nonrenewable resources are available in fixed amounts and recycling converts fixed to infinite

VOCABULARY DRILL ANSWERS					
Term	**Definition**	**Example (s)**	**Term**	**Definition**	**Example (s)**
anaerobic	x	t	malnutrition	u	q
atoms	b	b	matter	a	a
cellulose	v	r	mineral	f	f
chemical energy	n	l	mixture	e	e
compound	d	d	molecule	c	c, d
elements	b	b	natural organics	i	d
energy	k	i	organic molecule	g	d
entropy	q	o	organic phosphate	y	u
fermentation	w	s	overgrazing	z	v
First Law of Thermodynamics	o	m	oxidation	t	i, j, p, s, w
inorganic molecule	h	g	potential energy	m	h, k, l, r, s
kinetic energy	l	i, j	Second Law of Thermodynamics	p, r	n
Law of Conservation of Matter	aa	w	synthetic organics	j	l

CHAPTER 4

ECOSYSTEMS: POPULATIONS AND SUCCESSION

Population Dynamics

1. A balance between births and deaths is called a population ____________________.

Population Growth Curves

2. A population that continues to grow exponentially until it exhausts essential resources and then has a precipitous die-off demonstrates the __________ curve of population growth.

3. A population that grows exponentially and then exhibits a leveling off demonstrates the __________ curve of population growth.

4. Animal populations subjected to human disturbances usually demonstrate the __________ curve of population growth.

5. The _______ curve of population growth is an indicator of ecosystem nonsustainability.

Biotic Potential versus Environmental Resistance

6. Indicate whether the following factors are examples of biotic potential [BP] or environmental resistance [ER].

a. [ ] adverse weather conditions
b. [ ] ability to cope with weather conditions
c. [ ] predators
d. [ ] defense mechanisms
e. [ ] competitors
f. [ ] reproductive rate

7. If biotic potential is higher than environmental resistance, the population will (increase, decrease).

8. Population explosions occur when biotic potential is (higher, lower) than environmental resistance.

9. When a population is more or less stable in terms of numbers, biotic potential is (higher than, lower than, equal to) environmental resistance.

10. (True or False) There is a finite number of plants or animals that can be supported by an ecosystem.

11. Describe two types of reproductive strategies used by plants or animals.

a. ____

b. ____

12. (True or False) The dynamic balances which exist between biotic potential and environmental resistance may be easily upset by humans.

13. (True or False) When balances are upset, there are population explosions of some species and extinctions of other species.

Density Dependence and Critical Numbers

14. Environmental resistance (increases, decreases) as population density increases.

15. How are the concepts of critical numbers and endangered species related?

____

Mechanisms of Population Equilibrium

Predator-Prey Dynamics

16. Explain how Fig. 4-5 represents an example of predator-prey balance.

____

17. If a herbivore population is not controlled by natural enemies, it is likely that the herbivore population will (increase, decrease) and the vegetation it feeds on will (increase, decrease).

18. Which of the following groups of organisms requires a high degree of adaptation to maintain a balance between environmental resistance and biotic potential? (Check all that apply.)

[ ] predator [ ] prey [ ] parasite [ ] host

19. (True or False) A natural enemy can be so effective that it causes the extinction of its prey or host. If you said True, give an example. ____________________

20. If individuals are stressed by environmental resistance, indicate whether they would [+] or would not [-] exhibit the following characteristics.

[ ] less able to compete successfully with other species
[ ] more vulnerable to attack by predators
[ ] more vulnerable to attack by parasitic and disease organisms
[ ] more likely to attempt to disperse from the area

Plant-Herbivore Dynamics

21. Indicate whether the following activities will maintain [+] or upset [-] the plant-herbivore balance.

[ ] introducing goats on islands
[ ] killing off herbivore predators
[ ] confining herbivores in a localized area
[ ] allowing the herbivores to exceed the carrying capacity of the habitat
[ ] perpetuating monocultures

22. In undisturbed ecosystems, herbivore populations are held in check by predators and parasites. How does this demonstrate the Third Principles of Ecosystem Sustainability?

____

____

Competition

23. Identify three factors that prevent one plant species from driving out others in competion.

a. ____

b. ______________________________

c. ______________________________

24. Indicate whether the following conditions are [+] or not [-] examples of balanced herbivory.

[ ] plant species that demonstrate unique adaptations to specific microclimates
[ ] introducing plant species that outcompete native species
[ ] introducing herbivores that outcompete native grazing animals
[ ] introducing carnivores that outcompete native carnivores
[ ] introducing carnivores that prey species do not recognize
[ ] development of rubber plantations

25. (True or False) Effective territorial behavior by an animal results in priority use of those resource available in a specific area.

26. List three options for survival available to individuals unable to claim a territory.

a. ______________ b. ______________

c. ______________

Introduced Species

27. Explain how ecosystems got out of balance with the following introductions of exotics.

a. rabbits in Australia: ______________

b. chestnut blight in America: ______________

c. domestic cats on islands: ______________

d. zebra mussels in Great Lakes: ______________

28. Give an example of an introduction of an exotic that did not affect the balance.

____

29. List the two conditions that enable introduced species to thrive in their new ecosystems.

a. ______________________________

b. ______________________________

Disturbance and Succession

30. (True or False) Changes in ecosystem structure and/or function, over time, are inevitable.

31. How is your answer to question 29 consistent with or a contradiction to the Equillibrium Theory?

____

____

Ecological Succession

32. (True or False) There is an end point in the process of ecological succession?

33. List three examples of ecological succession.

a. ____________ b. ____________ c. ____________

34. In what order (1 to 4) would the following plant communities appear during primary succession?

[ ] trees [ ] mosses [ ] shrubs [ ] larger plants

35. How do mosses change the conditions of the area so that it can support larger plants?

____

36. Suppose that a section of eastern deciduous forest was cleared for agriculture and later abandoned. In what order (1 to 4) would the following species invade the area?

[ ] deciduous trees [ ] crabgrass [ ] pine trees [ ] grasses

37. What prevents hardwood species of the deciduous forest from reinvading the area immediately?

____

38. The end result of aquatic succession is a ____________ ecosystem.

Disturbance and Resilience

39. What role does disturbance play in promoting biodiversity?

____

40. In which of the following ecosystems can fire be used as a management tool? (Check all that apply.)

[ ] grasslands [ ] tropical rain forests [ ] redwood trees [ ] pine forests

41. Indicate whether the following events are [+] or are not [-] the results of controlled ground fires.

[ ] removal of dead litter
[ ] permanent harm of mature trees
[ ] seed release of some tree species

[ ] destruction of wood-boring insects
[ ] crown fires
[ ] loss of human lives and property

42. List three ways disturbances contribute to ecosystem structure.

a. ____________________ b. ____________________ c. ____________________

43. A ____________________ ecosystem is one that maintains its normal functioning even through a disturbance.

44. List four resilience mechanisms that become operational after fires.

a. ______________________________ b. ______________________________

c. ______________________________ d. ______________________________

45. Write the Fourth Principle of Ecosystem Sustainability.

____

46. Write the Fifth Principle of Ecosystem Sustainability.

____

Implications for Humans

47. List two ways humans can create a sustainable future.

a. ____

b. ____

48. The management paradigm that recognizes that some degree of management is necessary to maintain normal ecosystem structure and function where there are human demands on ecosystems is called ____________________ management.

49. List the four principles of ecosystem science.

a. ____

b. ____

c. ____

d. ____

50. The human population is exploding because we have effectively

a. increased our biotic potential (True or False)
b. decreased our environmental resistance (True or False)

51. The human population growth pattern presently resembles a _________ curve.

52. What is the carrying capacity of our planet for human beings?

a. 6 billion   b. 10 billion   c. unlimited   d. cannot be determined

VOCABULARY DRILL

Directions: Match each term with the list of definitions and examples given in the tables below. Term definitions and examples may be used more than once.

Term	Definition	Example(s)	Term	Definition	Example(s)
aquatic succession			host specific		
balanced herbivory			monoculture		
biotic potential			natural enemies		
carrying capacity			population explosion		
climax ecosystem			population density		
critical number			primary succession		
ecological restoration			recruitment		
ecological succession			replacement level		
environmental resistance			reproductive strategies		
exponential increase			secondary succession		
fire climax			territoriality		

VOCABULARY DEFINITIONS
a. number of offspring a species may reproduce under ideal conditions
b. number of offspring that survive and reproduce
c. the way species achieve recruitment in their populations
d. a doubling of each generation
e. the result of exponential increase
f. biotic and abiotic factors that tend to decrease population numbers

VOCABULARY DEFINITIONS
g. number of offspring required to replace breeding population
h. number of individuals per unit area
i. the minimum population base (number) of a species
j. predators and parasites that control population size
k. defending territory against encroachment of others of the same species
l. maximum population a given habitat will support without the habitat being degraded over time
m. growth of a single species over a wide area
n. parasites that attack only one species and close relatives
o. balance among competing plant populations maintained by herbivores
p. the orderly transition from one biotic community to another
q. a stage of equilibrium between all species and the physical environment
r. succession in an area that has not been previously occupied
s. succession in an area that has been previously occupied
t. replacement of aquatic by terrestrial communities
u. ecosystems that depend on the recurrence of fire to maintain existing balance
v. reclaiming natural ecosystems

VOCABULARY EXAMPLES
a. live births, eggs laid, seeds in plants
b. seeds that sprout and produce more seeds
c. produce massive numbers of young
d. 2 → 4 → 8 → 16 → 32
e. lack of food, space, and mates
f. number of offspring required to replace breeding population
g. 2 deer/acre
h. endangered species
i. predators and parasites
j. scent marking
k. upper limit on S-shaped population growth curve
l. a field of cotton
m. host specific: dog tapeworm

VOCABULARY EXAMPLES
n. the opposite of monoculture
o. bare rock → moss → herbaceous plants → shrubs → trees
p. deciduous forest
q. deciduous forest → monoculture → grassland → shrubs → pine forest → deciduous forest
r. lake → marsh → grassland
s. grasslands and pine forests
t. wetlands

SELF-TEST

Indicate whether the following factors are examples of biotic potential [BP] or environmental resistance [ER].

1. [ ] lack of suitable habitat
2. [ ] disease
3. [ ] reproductive rate
4. [ ] ability to migrate
5. [ ] + adaptability
6. [ ] + competition

Identify whether the following traits are characteristic of [J] or [S] population growth curves.

7. [ ] biotic potential = environmental resistance

8. [ ] biotic potential > environmental resistance

9. [ ] population increases and dies off in rapid sequence

10. [ ] violates the Fourth Principle of Ecosystem Sustainability

11. [ ] the human population growth curve is an example

12. What concept concerning population numbers is the direct opposite of carrying capacity?

    ____

13. In what order (1 to 4) would the following plant communities appear during primary succession?

    [ ] trees [ ] mosses [ ] grasses [ ] shrubs

14. In what order (1 to 4) would the following plant communities appear during secondary succession?

    [ ] grasses [ ] pine trees [ ] herbaceous plants [ ] deciduous trees

15. What ecological factors define primary versus secondary succession?

____

16. In which of the following ecosystems can fire not be used as a management tool?

a. grasslands   b. coniferous forests   c. hardwood forests   d. both a and b

17. Which of the following is not an intended function of ground fires?

a. removal of dead litter   b. seed release of some tree species
c. starting crown fires   d. destruction of wood-boring insects

18. Aggressive defense of habitat is called

a. biotic potential   b. succession   c. territoriality   d. the niche

19. The process whereby the growth of one community causes changes that make the environment more favorable to a second community and less favorable to the first is called

a. biotic potential   b. succession   c. territoriality   d. the niche

20. Which of the following is not a characteristic of climax ecosystems?

a. have maximum species diversity
b. have different dominant plant forms
c. are self-perpetuating
d. will maintain themselves regardless of climatic, abiotic, or biotic changes

Which of the following human actions will [+] and will not [-] cause upsets of natural ecosystems? (Note: The following list expands on that given in the text.)

21. [ ] increasing population size

22. [ ] decreasing the gross national product

23. [ ] supporting conservation organizations

24. [ ] introducing species from one ecosystem to another

25. [ ] eliminating natural predators

ANSWERS TO STUDY GUIDE QUESTIONS

1. equilibrium; 2. J; 3. S; 4. J; 5. J; 6. ER, BP, ER, BP, ER, BP; 7. increase; 8. higher; 9. equal to; 10. true; 11. have massive numbers of offspring, low reproductive rate with long-term care of young; 12. true; 13. true; 14. increases; 15. Species become endangered when they reach critical numbers; 16. figure shows predictable rises and falls in populations of wolves and moose; 17. increase, decrease; 18. all should be checked; 19. true, humans; 20. all +; 21. all -; 22. The size of consumer populations in ecosystems is maintained such that overgrazing and other forms of overuse do not occur; 23. microclimates or microhabitats, single species cannot use all the resources in the area, balanced herbivory; 24. +,-,-,-,-,-; 25. true; 26. live to maturity, die, migrate; 27. rabbit population exploded and overgrazed grassland, wiped out American chestnut, overly effective predators, endemic species displacement; 28. ring-necked pheasant; 29. lack of competition with endemic species, no predators; 30. true; 31. It is not consistent because nothing stays the same forever; 32. true; 33. primary, secondary, aquatic; 34. 4, 1, 3, 2; 35. provide first layer of humus; 36. 4, 1, 3, 2; 37. temperature too hot; 38. terrestrial; 39. provides populations for invasions of new habitats; 40. grasslands, redwoods, pine forests; 41. +,-,+,+,-,-; 42. remove organisms, reduce populations, create opportunities for other species to colonize; 43. resilient; 44. See Fig. 4-21; 45. Ecosystems show resilience during disturbance; 46. Ecosystems depend on diversity; 47. Protecting and managing the natural environment to maintain the goods and services vital to human economy and survival. Establishing a balance between our own species and the rest of the biosphere; 48. adaptive; 49. ecosystems will change over time, individual species and interactions between species will have important impacts on ecological processes, different sites and regions are unique (abiotic and biotic characteristics), disturbances are important and frequent events that have a profound impact on the ecosystem; 50. true, true; 51. J; 52. d

---

ANSWERS TO SELF-TEST

1. ER; 2. ER; 3. BP; 4. BP; 5. BP; 6. ER; 7. S; 8. J; 9. J; 10. J; 11. J; 12. critical number; 13. 4, 1, 3, 2; 14. 1, 3, 2, 4; 15. a suitable soil base to start with; 16. c; 17. c; 18. c; 19. b; 20. d; 21. -; 22. +; 23. +; 24. -; 25. -

VOCABULARY DRILL ANSWERS					
Term	**Definition**	**Example(s)**	**Term**	**Definition**	**Example(s)**
aquatic succession	t	r	host specific	n	m
balanced herbivory	o	n	monoculture	m	l
biotic potential	a	a	natural enemies	j	i, m
carrying capacity	l	k	population explosion	e	d
climax ecosystem	q	p	population density	h	g
critical number	i	h	primary succession	r	o
ecological restoration	v	t	recruitment	b	b
ecological succession	p	o, q, r	replacement level	g	f
environmental resistance	f	e	reproductive strategies	c	c
exponential increase	d	d	secondary succession	s	q, r
fire climax	u	s	territoriality	k	j

CHAPTER 5

ECOSYSTEMS: AN EVOLUTIONARY PERSPECTIVE

Selection by the Environment

1. Identify two ways in which changes in the gene pool might occur.

   a. ______________________ b. ______________________

2. (True or False) All individuals in a population reproduce at the same rate.

3. (True or False) Survival automatically leads to reproduction.

Change Through Selective Breeding

4. Selective breeding requires a (repetitive, one-time) process in order to develop the desired trait.

5. Selective breeding (a. increases, b. decreases, c. both a and b) certain alleles in the gene pool of a population.

6. All the different breeds of dogs are derived from the (same, different) wild dog gene pool.

7. (True or False) It is genetically logical to mate a Great Dane with a Chihuahua.

Change Through Natural Selection

8. In nature, every generation of every species is subjected to an intense selection of ________________ and ________________.

9. New alleles that provide for or enhance a trait that aids in survival and reproduction will (increase, decrease) in the gene pool of the population.

Adaptations to the Environment

10. Indicate whether the following adaptations would [+] or would not [-] support the survival and reproduction of organisms.

    a. [ ] coping with climatic and other abiotic factors
    b. [ ] obtaining food and water
    c. [ ] escaping from predation
    d. [ ] ability to attract mates
    e. [ ] ability to migrate to adjacent areas

11. Give an example of adaptations a to e in question 10. (See Fig. 5-4.)

    a. ____

    b. ____

    c. ____

    d. ____

    e. ____

Selection of Traits and Genes

12. Provide an example of a trait related to:

    a. physical appearance ________________________________

    b. tolerance ________________________________

    c. behavior ________________________________

    d. metabolism ________________________________

13. The hereditary component of all traits of all organisms is encoded in ______________ molecules.

14. The collection of all the DNA molecules or genes within a cell of an organism is called its __________________ makeup.

15. Relate DNA to structural, enzymatic, and hormonal proteins to physical, metabolic, and growth traits in the diagram below. (See Fig. 5-5 for an example of this exercise.)

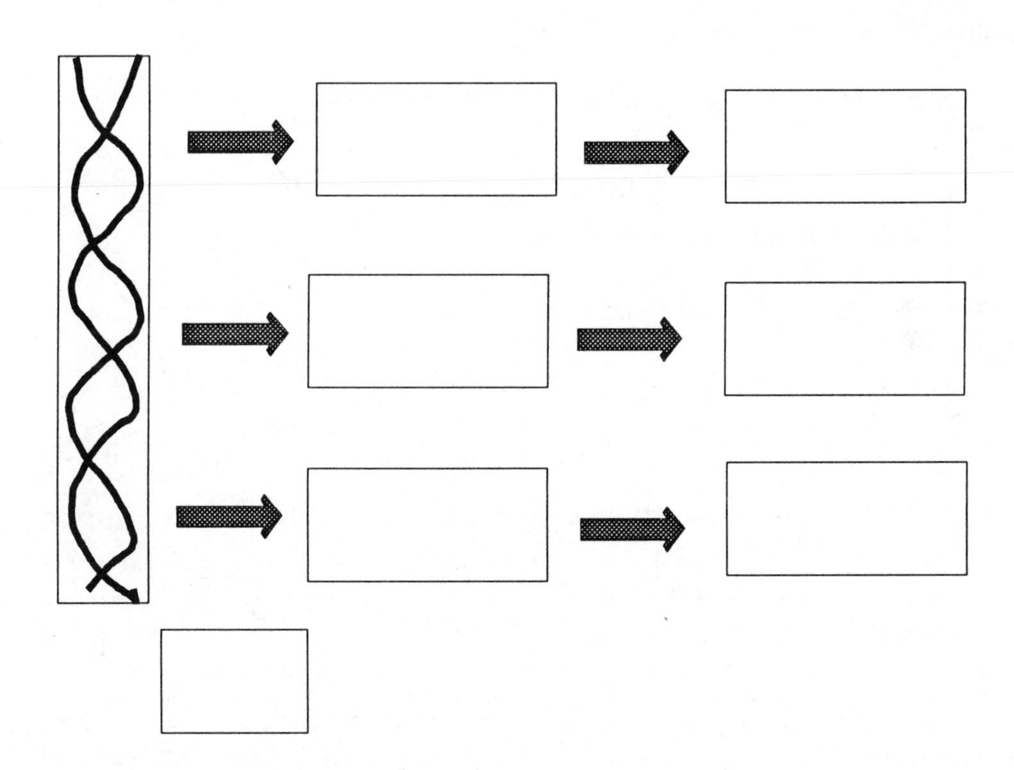

Genetic Variation and Gene Pools

16. The genetic makeup of nearly all organisms consists of (one, two, three, four) complete set(s) of genes or (one, two, three, four) pair(s).

17. Indicate whether each of the following cell types would have one [1] or two [2] alleles of a gene pair.

[ ] body cells [ ] sperm cell [ ] egg (ovum) [ ] fertilized egg

18. A sperm cell entering an egg is called the process of ________________.

19. Each fertilization will result in a different combination of ____________________ and this combination will be different from that of either parent.

20. Explain why it is more likely for males to contract sex-chromosome-related diseases than females.

____

21. What is the genetic difference between identical twins and clones?

____

Mutations: The Source of New Alleles

22. Explain what a mutation is in terms of the DNA molecule.

____

23. (True or False) Mutations are rarely beneficial.

Changes in Species and Ecosystems

24. Natural selection on gene pools is a (quick, slow) process.

Limits of Change

25. List the three options available to each species that is ill-adapted to a new ecosystem.

a. __________________ b. __________________ c. __________________

26. Adaptations are:

a. present in a species gene pool prior to environmental selection
b. the result of mutations occurring after the environment selects for a particular trait
c. variations on preexisting genetic themes in the species gene pool
d. Both a and c are correct.

27. What two criteria must be met in order for survivors to adapt to a new environmental condition?

a. ______________________ b. ______________________

28. Indicate whether the following are [+] or are not [-] criteria for populations of individuals to survive new conditions.

[ ] high degree of genetic variation in the gene pool
[ ] wide geographic distribution
[ ] high reproductive capacity
[ ] being small

29. Compare cockroaches and elephants in terms of whether they do [+] or do not [-] meet the following criteria for survival.

Cockroaches

[ ] high degree of genetic variation in the gene pool
[ ] wide geographic distribution
[ ] high reproductive capacity
[ ] being small

Elephants

[ ] high degree of genetic variation in the gene pool
[ ] wide geographic distribution
[ ] high reproductive capacity
[ ] being small

30. (Cockroaches or Elephants) have the best chance of adapting to new environments.

31. (Cockroaches or Elephants) are more likely to adapt to changes caused by humans.

Speciation

32. Rank (1 to 4) the following events in terms of their order of occurrence in the process of speciation.

[ ] two or more species developing from one
[ ] original species gene pool separated
[ ] long-term loss of interbreeding potential
[ ] single species gene pool

33. If a species moves to a new location where biotic and/or abiotic conditions are somewhat different, which of the following events could happen?

a The migrant population may die out. (True or False)
b. Some individuals may survive and reproduce. (True or False)
c. There will be selection pressure for alleles that enhance survival. (True or False)

d. Alleles in the gene pool of the migrant population will become different from those in the gene pool of the original population. (True or False)
e. Speciation could occur. (True or False)

34. If natural selective pressures are for animals with long hair and short legs, then the majority of the population will have (long, short) hair and (long, short) legs.

35. Give one example of migration and speciation that occurred on the Galapagos Islands.

____

36. Speciation

a. is a rapid process
b. is a slow process
c. can be rapid or slow, depending on the species and environmental conditions

Evolving Ecosystems?

37. (All, A few) species within an ecosystem undergo simultaneous adaptation and change through natural selection.

38. Every species exists and can only exist in the context of:

a. producers b. consumers c. decomposers d. the whole ecosystem

39. Identify the original organisms that occupied the grazing niche:

a. on Galapagos Islands: ___________________________________

b. in Brazil: ___________________________________

c. in North America: ___________________________________

d. in Australia: ___________________________________

40. Explain why the natural selection process chose different species to occupy the grazing niche in each of the areas listed above.

____

41. (True or False) If you introduced rabbits into Australia, they would become the prime herbivores.

42. (True or False) If you introduced cattle into Australia, they would become the prime herbivores.

Plate Tectonics

43. The Earth's crust is (thick, thin) and in a constant state of (movement, inertia).

Tectonic Plates

44. The floating slabs of rock that make up Earth's crust are called ________________ plates.

45. Define Pangea: ____

46. How fast are the tectonic plates moving? __________ cm/year.

Match the type of tectonic plate interaction with the resultant outcome.

Tectonic Plate Interaction	Resultant Outcomes
47. divergent boundaries	a. mountains
48. transform boundaries	b. mid-ocean ridges
49. convergent boundaries	c. 1994 Los Angeles earthquake
50. collision	d. 1980 eruption of Mount Saint Helens

51. List three ways that movement of Earth's crust affects climate.

a. ____

b. ____

c. ____

Evolution in Perspective

The Fossil Record

52. If the entire geological history of Earth was made equivalent to one year, when did humans:

a. appear? ____________________ b. develop agriculture? ____________________

c. develop technological knowledge? ____________________

Controversy over Evolution

53. List three major evidences that evolution is the process that accounts for the present array of living organisms.

a. ____________________ b. ____________________ c. ____________________

54. Which of the following are [+] and are not [-] indicators that humans are going to determine the future course of evolution of life on Earth?

[ ] altered the surface of the Earth
[ ] caused the extinction of numerous species
[ ] moved many species between continents
[ ] humans are the only species able to exercise conscious choice

Stewardship of Life

55. Biodiversity is an expression of __________________ variation in gene pools.

56. Reducing the size of surviving populations reduces __________________ variation in gene pools.

57. The inability to adapt to new conditions is the result of the lack of __________________ variation in gene pools.

58. Relate the Fourth Principle of Ecosystem Sustainability to:

a. endangered species: ____

b. agriculture: ____

c. biotechnology: ____

d. medicine: ____

VOCABULARY DRILL

Directions: Match each term with the list of definitions and examples given in the tables below. Term definitions and examples might be used more than once.

Term	Definition	Example(s)	Term	Definition	Example(s)
adaptation			mutation		
alleles			natural selection		
biological evolution			selective breeding		
differential reproduction			selective pressures		
DNA			speciation		
gene			tectonic plates		
gene pool			trait		
genetic variation					

VOCABULARY DEFINITIONS
a. total genetic differences that exist among individuals within a population
b. total number of genes within a population
c. individuals reproducing more than others in a population
d. change in the gene pools of species over the course of generations
e. a human-induced form of differential reproduction
f. environmental resistance factors that determine survival and reproductive rates
g. selection and modification of species' gene pools through natural forces
h. any particular characteristic of an individual
i. a process that results in two or more species from one
j. huge slabs of rock floating on Earth's molten core
k. the ability to form balanced relationships with biotic community and abiotic environment
l. the blueprint of the total genetic makeup of organisms
m. a segment of DNA that codes for one particular protein
n. different forms of a gene
o. inheritable change in the DNA molecule

VOCABULARY EXAMPLES	
a. races of humans	g. Galapagos finches
b. 50 to 100 K for humans	h. San Andreas fault
c. Great Danes and dachshunds	i. a giraffe's long neck
d. fish, amphibians, reptiles, mammals	j. double alpha-helix
e. every factor of environmental resistance	k. nucleotide sequence
f. blue eyes, dark skin	l. altered nucleotide sequence

SELF-TEST

Indicate whether the following terms or descriptions refer to traits [T] or genes [G].

1. [ ] DNA segment
2. [ ] physical appearance
3. [ ] alleles

Check any of the following events that increase genetic diversity in a species gene pool.

4. [ ] sexual reproduction
5. [ ] selective breeding
6. [ ] recombination

7. Which of the following statements is not true concerning DNA molecules? DNA molecules are:

a. the site of genetic information within organisms
b. found only in sperm or egg cells
c. capable of being influenced by genetic engineering
d. one of the locations for mutations to occur

8. How many alleles of a gene pair would normally be found in a sperm cell?

a. one b. two c. three d. four

9. How many alleles of a gene pair would normally be found in a skin cell?

a. one b. two c. three d. four

Complete each of the following statements with the terms MORE or LESS.

10. More selective pressure leads to ________________ reproduction.

11. Less adaptation to abiotic and biotic factors leads to _______________ reproduction.

12. Explain why more survival does not automatically lead to more reproduction.

____

____

13. Which of the following statements is not correct regarding selective breeding?

a. It is a one-time event.
b. It increases genetic diversity in the gene pool of a population.
c. It decreases genetic diversity in the gene pool of a population.
d. It is a human-induced mechanism to control natural selection.

14. Mutations that are passed on from generation to generation are called:

a. beneficial b. neutral c. lethal d. both a and b

15. Explain why the giant tortoises were so successful in occupying the herbivore niche on the Galapagos Islands.

____

____

Which of the following adaptations gave goats the greatest advantage over giant tortoises as the prime herbivores on the Galapagos Islands. (Check all that apply and defend selections.)

	Adaptations	Defense of Answer
16.	[ ] coping with abiotic factors	______________________
17.	[ ] obtaining food	______________________
18.	[ ] escaping from predation	______________________
19.	[ ] ability to migrate	______________________

Identify the criteria for species survival in the list below and then indicate whether each is a criterion of the human population. (Check all that apply.)

	Criteria for Survival	Present in Human Population
20.	[ ] genetic diversity in gene pool	23. [ ]
21.	[ ] wide geographic distribution	24. [ ]

22. [ ] rapid reproductive rate 25. [ ]

Identify those conditions that represent the human system (check all that apply) and then predict whether the listed conditions would promote [+] or prevent [-] the future survivability of the human species.

	Conditions	Survivability
26.	[ ] causing rapid abiotic/biotic changes	30. [ ]
27.	[ ] monoculture development	31. [ ]
28.	[ ] sustainable agriculture	32. [ ]
29.	[ ] preserving biodiversity	33. [ ]

34. Explain the contradiction you should have noticed in your responses to questions 20 to 25 and questions 26 to 33.

____

____

35. Identify a major force behind climate changes: ______________________________

ANSWERS TO STUDY GUIDE QUESTIONS

1. selective breeding, natural selection; 2. false; 3. false; 4. repetitive; 5. c; 6. same; 7. true; 8. survival, reproduction; 9. increase; 10. all +; 11. see Fig. 5-4; 12. the way an organism looks, athletic ability, gentle or aggressive, sensitivity to allergens; 13. DNA; 14. genetic; 15. see Fig. 5-5; 16. two, one; 17. 2, 1, 1, 2; 18. fertilization; 19. genes or alleles; 20. males have only one X chromosome that may or may not carry the allele can never be safe in a heterozygous condition; 21. twins are the result of sexual reproduction whereas clones are the result of asexual reproduction; 22. change in nucleotide sequence; 23. true; 24. slow; 25. adapt, move, die; 26. d; 27. some, enough; 28. all plus; 29. cockroaches (all +), elephants (all -); 30. cockroaches; 31. cockroaches; 32. 4, 2, 3, 1; 33. all true; 34. long, short; 35. finches; 36. c; 37. all; 38. d;. 39. giant tortoise, termites, bison, kangaroo; 40. Nature can only modify preexisting species; 41. true; 42. true; 43. thin, movement; 44. tectonic; 45. The original continental landmass that separated into the major continents as they are presently known; 46. 6; 47. b; 48. c; 49. d; 50. a; 51. movement of continents to different positions on the globe alters climates AND direction and flow of ocean currents, uplifting of mountains alters movement of air currents; 52. 3 p.m. on December 31, last 2 minutes of the year, last 2 seconds of the year; 53. fossil record, species evolution, DNA nucleotide similarities and differences; 54. all +; 55. genetic; 56. genetic; 57. genetic; 58. small gene pool with little variability to fall back on, no original genetic stocks to go back to, no source of genes to transfer, one-quarter of all prescription drugs have active ingredients developed from wild plants

---

ANSWERS TO SELF-TEST

1. G; 2. T; 3. G; 4. check; 5. blank; 6. check; 7. b; 8. a; 9. b; 10. LESS; 11. LESS; 12. Not all organisms that survive have the physical, behavioral, and other adaptations necessary to reproduce; 13. b; 14. d; 15. The terrestrial herbivore niche was unoccupied; 16 to19. The greatest advantage goats had over giant tortoises was probably their ability to obtain food (i.e., goats are more effective herbivores), goats are also faster on land than tortoises, which gave them an additional advantage of escaping from predators; 20 to 25. all checked; 26. check; 27. check; 28. blank; 29. blank; 30. -; 31. -: 32. +; 33. +; 34. Humans have the ecological criteria for survival but have introduced conditions that make the ecosystem nonsustainable for them and all other species; 35. movement of Earth's crust (i.e., tectonic plates)

VOCABULARY DRILL ANSWERS

Term	Definition	Example(s)	Term	Definition	Example(s)
adaptation	k	i	mutation	o	l
alleles	n	k	natural selection	g	e
biological evolution	d	d	selective breeding	e	c
differential reproduction	c	c	selective pressures	f	e
DNA	l	j	speciation	i	g
gene	m	k	tectonic plates	j	h
gene pool	b	b	trait	h	f
genetic variation	a	a			

CHAPTER 6

THE HUMAN POPULATION: DEMOGRAPHICS

The Population Explosion and Its Cause

The Explosion

1. Chart the progress of human population growth from 1830 to that projected for 2046.

Year	Billions of People	Years to Reach Population
a. 1830	______________	____________________

Year	Billions of People	Years to Reach Population
b. 1930	________	________
c. 1960	________	________
d. 1975	________	________
e. 1987	________	________
f. 1998	________	________
g. 2009	________	________
h. 2020	________	________
i. 2033	________	________
j. 2046	________	________

Reasons for the Explosion

2. Why was the human population in a dynamic balance prior to the 1800s?

____

3. List four factors that increased the recruitment rate of the human population after the 1800s.

a. ____________________ b. ____________________

c. ____________________ d. ____________________

4. Human population growth is expected to level off at ________ billion.

Different Worlds

Rich Nations and Poor Nations

5. The countries of the world fall into three major economic categories (a. high-income, highly developed; b. middle-income, moderately developed; and c. low-income). Identify these categories with examples of countries or regions that are in each category (see Fig. 6-3).

[ ] United States, Japan, Western Europe
[ ] Latin America, East Asia
[ ] East and Central Africa, India, China

6. Developed countries have about ________ percent of Earth's population, but they control about ________ percent of the world's wealth.

7. Developing countries have ________ percent of Earth's population, but control only ________ percent of world's wealth.

8. List six conditions of absolute poverty.

a. ____________ b. ____________ c. ____________

d. ____________________ e. ____________________ f. ____________________

Population Growth in Rich and Poor Nations

9. Identify the countries that have the shortest and longest doubling times (see Table 6-3).

   a. ____________________ b. ____________________

10. Over the past few decades, the fertility rates in developed countries have (Increased, Decreased) and in developing countries have (Increased, Decreased).

Different Populations Present Different Problems

11. Match the characteristics below with [a] developed or [b] developing countries.

   [ ] high total fertility rates
   [ ] low total fertility rates
   [ ] short doubling times
   [ ] high consumptive lifestyles
   [ ] long doubling times
   [ ] annual rates of increase $\geq$ 2%
   [ ] annual rates of increase $<$ 1%
   [ ] population growth is most intense
   [ ] poverty is most intense
   [ ] environmental degradation is most intense
   [ ] eating high on the food chain
   [ ] demand for tropical hardwoods
   [ ] demand for exotic wildlife

12. The negative impacts of high consumptive lifestyles may be moderated by a factor called ____________________ regard.

13. What three conditions must be met in order to achieve a sustainable population?

   a. ____________________ b. ____________________

   c. ____________________

Environmental and Social Impacts of Growing Populations and Affluence

The Growing Populations of Developing Nations

14. List five ways people in developing nations are attempting to resolve population growth and land availability.

   a. ____________________ b. ____________________

   c. ____________________ d. ____________________

   e. ____________________

15. Indicate whether the following are [+] or are not [-] environmental conditions resulting from attempts to increase agriculture in developing countries.

[ ] decrease in per capita agricultural land
[ ] attempt agriculture on marginal land
[ ] soil erosion
[ ] loss of topsoil
[ ] water pollution
[ ] loss of soil productivity
[ ] deforestation

16. What is the fundamental problem with attempting to open new lands for agriculture in developing countries? ____

17. Indicate whether the following are [+] or are not [-] environmental/social conditions resulting from people in developing countries attempting to seek a better life by moving from the country to cities.

[ ] high population densities
[ ] lack of utilities and social services
[ ] polluted water
[ ] disease
[ ] high crime rates

18. Identify two sources of illicit income that people in developing countries might turn to in order to earn enough money to survive.

a. ________________________ b. ________________________

19. The majority of customers for the products of illicit income come from (developed, developing) countries.

20. Many people in developing countries see (emigration, immigration) as their best hope for a brighter future.

21. Developed countries are placing (more, fewer) restrictions on immigration.

22. (True or False) Women in developing countries have access to an adequate social welfare system.

23. (True or False) Cities in developing countries have high numbers of stray or abandoned children.

Effects of Increasing Affluence

24. (True or False) The effects of affluence are all negative.

25. The United States contains _________ percent of Earth's population and is responsible for _________ percent of the global emissions of carbon dioxide.

26. Explain why the cities in developing countries are more polluted (air and water) than cities in developed countries.

____

27. Take a stand! What is your position [a. agree, b. neutral, c. disagree] concerning the following population issues?

[ ] Further population growth will be beneficial.
[ ] Our technology will solve any problems associated with population growth.
[ ] Even at a population of 12 billion, we can all enjoy an affluent lifestyle.

Dynamics of Population Growth

Population Profiles

28. Population profiles can be used to project future:

a. ______________________________ b. ______________________________

29. Indicate whether the following fertility rates will lead to an increase [+], decrease [-], or stability [0] in population growth.

[ ] fertility rate < 2
[ ] fertility rate = 2
[ ] fertility rate > 2

30. Which of the above fertility rates is known as replacement fertility? __________

31. Match the following shapes of population profiles with the fertility rates given in question 29.

a. column-shaped: Fertility rate = ________
b. pyramid-shaped: Fertility rate = ________
c. upside-down pyramid: Fertility rate = ________

32. The population pyramid of Denmark would be __________________ shaped.

33. The population pyramid of Kenya would be __________________ shaped.

Population Projections

34. Identify two population projections provided in population profiles.

a. ______________________________ b. ______________________________

Based on the population profiles for Italy and Iraq in 2025, predict whether MORE or LESS of the following goods and services will be needed over the next 20 years.

Service	Italy	Iraq
schools	35. ________	39. ________
facilities for old people	36. ________	40. ________
new housing units	37. ________	41. ________
utility services	38. ________	42. ________

Population Momentum

43. Given the population momentum of Iraq, its population will continue to grow for __________ to __________ years after achieving replacement fertility levels.

The Demographic Transition

44. What has been the observed trend in economic development and fertility rates?

____

45. Subtracting the crude death rate from the crude birthrate (CBR-CDR) and then dividing by 10 gives the annual rate of __________________________.

46. Doubling time for a population may be found by dividing the annual rate of increase into (7, 70, 700).

47. Calculate the annual rate of increase and doubling time for the following countries.				
Country	Crude Birthrate	Crude Death Rate	Rate of Increase	Doubling Time
Kenya	46	7		
Mexico	29	6		
United States	17	9		
Denmark	12	12		

48. The goal of each country should be to have a (long, short) doubling time.

49. The demographic transition refers to changes in infant __________________ and __________________ rates, which tend to parallel development.

50. There are (one, two, three, four) major phases in the demographic transition?

51. In the table below, indicate whether the variables given after each phase are high [+], declining [-], or stable [0].

Phase	Birthrate	Death Rate	Population
I	[ ]	[ ]	[ ]
II	[ ]	[ ]	[ ]
III	[ ]	[ ]	[ ]
IV	[ ]	[ ]	[ ]

52. Developed nations are in phase __________ .

53. Developing nations are in phase __________ .

VOCABULARY DRILL

Directions: Match each term with the list of definitions and examples given in the tables below. Term definitions and examples might be used more than once.

Term	Definition	Example(s)	Term	Definition	Example(s)
absolute poverty			doubling time		
age structure			environmental regard		
crude birth rate			natural increase		
crude death rate			population profile		
demographer			population momentum		
demographic transition			replacement fertility		
developed countries			total fertility rate		
developing countries					

VOCABULARY DEFINITIONS
a. fertility rate that will just replace the population of parents
b. collecting, compiling, and presenting information regarding populations
c. continued population growth after fertility rate = replacement rate
d. the gradual shift from primitive to modern condition that is correlated with development
e. number of births per 1000 of the population per year
f. number of deaths per 1000 of the population per year
g. relative proportion of people in each age group
h. net increase or loss of population per year
i. the number of years it will take a population growing at a constant percent per year to double
j. factors that moderate environmental impacts of affluent lifestyles
k. average number of children each woman has over her lifetime
l. a condition of life so limited by malnutrition, illiteracy, disease, high infant mortality, and low life expectancy as to be beneath any definition of human decency
m. high-income nations
n. middle- to low income-nations
o. the number of people at each age for a given population

VOCABULARY EXAMPLES	
a. 90 percent of the people in developing countries	g. (CBR - CDR)) 10
b. Table 6-5	h. Fig. 6-10
c. Fig. 6-16	i. 50 to 60 years in Iraq
d. Iraq	j. 2.0
e. 18 to 1,155 years	k. 1.3 to 6.7
f. recycling + conservation + pollution control	l. United States

SELF-TEST

1. Draw a graph showing the change in human population size on the Y axis from historical times through the present and projected into the future on the X axis.

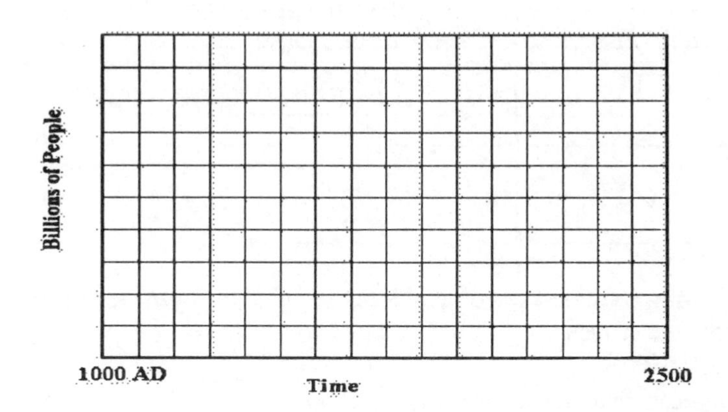

Identify the following countries or regions as [a] developed or [b] developing countries.

2. [ ] Latin America 3. [ ] Japan 4. [ ] Central Africa 5. [ ] United States

6. Developed countries have about _________ percent of Earth's population but control about _________ percent of the world's wealth.

7. Calculate the people-to-resource ratio for developed countries in units of resource per person. __________

8. Developing countries have about __________ percent of Earth's population but control only _________ percent of the world's wealth.

9. Calculate the people-to-resource ratio for developing countries in units of resource per person. __________

Assume that people living in developed countries reduced their resource consumption by 50 percent and donated the additional resource to developing countries.

10. Calculate the new people-to-resource ratio for developed countries in units of resource per person. __________

11. Calculate the new people-to-resource ratio for developing countries in units of resource per person. _________

Explain the results of your calculations in questions 10 and 11 in terms of:

12. Impact on developed countries: ____

13. Impact on developing countries: ____

Calculate the annual rate of increase and doubling time for the countries listed below.

Country	CBR	CDR	Rate of Increase	Doubling Time
14. Developed	14	9		
15. Developing	34	10		

Indicate whether the following consequences of exploding population are the results of high fertility rates in [a] developing countries, the affluence of [b] developed countries, or [c] both.

16. [ ] population pressure in the countryside

17. [ ] pressures on forests and endangered species

18. [ ] global pollution

19. [ ] high emigration rates

20. Give one example of how population profiles can be used.

____

21. The phase of demographic transition known as "modern stability" is in Phase (I, II, III, IV).

22. A population profile shows

a. standard of living

b. causes of death

c. numbers of people in each 5-year age group

d. factors that control fertility rates

23. The most significant factor(s) that determine how fast a population grows is (are)

a. age structure of the population

b. total fertility rate

c. infant and childhood mortality rates

d. all of the above

24. The population profile of a country with a "graying population" has the shape of a

a. column b. pyramid c. upside-down pyramid d. football on its point

25. Why would Iraq have to wait 50 years to experience a population decrease?

____

ANSWERS TO STUDY GUIDE QUESTIONS

1. 1, since the beginning of human history; 2, 100; 3, 30; 4, 15; 5, 12; 6, 11; 7, 11; 8, 11; 9, 13; 10, 13; 2. low recruitment; 3. cause of diseases recognized, improvements in nutrition, discovery of penicillin, improvements in medicine; 4. 12; 5. a, b, c; 6. 20, 80; 7. 80, 20; 8. malnutrition, illiteracy, disease, squalid surroundings, high infant mortality, low life expectancy; 9. Kenya, Italy; 10. decreased, increased; 11. b, a, b, a, a, b, a, b, b, b, a, a, a; 12. environmental; 13. stabilize population, decrease consumptive lifestyles, increase stewardly care; 14. subdivide farms, open new land for agriculture, move to cities, engage in illicit activities, emigrate to other countries; 15. all +; 16. no suitable land left to open for agriculture; 17. all +; 18. drugs, poaching; 19. developed; 20. emigration; 21. more; 22. false; 23. true; 24. false; 25. 4.5, 25; 26. Developed countries mine the resources of developing countries and leave their wastes in developing countries; 27. answer depends on your point of view; 28. population growth, demand for goods and services; 29. -, 0, +; 30. = 2; 31. = 2, > 2, < 2; 32. column; 33. pyramid; 34. births, deaths; 35. less; 36. more; 37. less; 38. less; 39. more; 40. less; 41. more; 42. more; 43. 50, 60; 44. inverse relationship between economic development and fertility rates; 45. increase; 46. 70; 47. see Table 6-5; 48. long; 49. births, deaths; 50. four; 51. +, +, 0; +, -, +; -, -, -; -, -, 0; 52. IV; 53. III

ANSWERS TO SELF-TEST

1. draw a J-shaped curve; 2. b; 3. a; 4. b; 5. a; 6. 20, 80; 7. 1:4; 8. 80, 20; 9. 1:<1; 10. 1:2; 11. 1:<1; 12. considerable change in quality of life; 13. very small change in resource availability; 14. 2.4, 30; 15. 0.5, 137; 16. L; 17. B; 18. B; 19. L; 20. make projections on future population trends; 21. IV; 22. c; 23. d; 24. a; 25. population momentum

VOCABULARY DRILL ANSWERS

Term	Definition	Example(s)	Term	Definition	Example(s)
absolute poverty	l	a	doubling time	i	e
age structure	g	h	environmental regard	j	f
crude birthrate	e	b	natural increase	h	g
crude death rate	f	b	population profile	o	h
demography	b	h	population momentum	c	I
demographic transition	d	c	replacement fertility	a	j
developed countries	m	l	total fertility rate	k	k
developing countries	n	d			

CHAPTER 7

ISSUES IN POPULATION AND DEVELOPMENT

STUDY QUESTIONS

Reassessing the Demographic Transition

1. Interpret the intersection of the lines for food production and population growth in terms of what Thomas Malthus predicted in 1798.

    ____

    ____

    ____

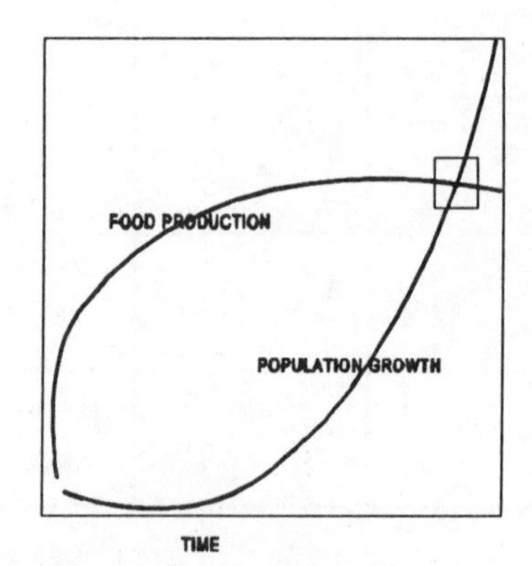

2. What two considerations were not included in the Malthusian prediction?

a. ______________________________ b. ______________________________

3. What are two basic schools of thought on the issue of population and development?

a. ______________________________ b. ______________________________

Factors Influencing Family Size

4. Indicate whether the following social and economic factors determine the family size of [a] developing, [b] developed, [c] both, or [d] neither countries.

[ ] children: an economic liability
[ ] old-age security
[ ] importance of education
[ ] status of women
[ ] religious beliefs
[ ] high infant and childhood mortality rates
[ ] use of contraceptives

Conclusions

5. Poverty leads to (more or less) environmental degradation, which leads to (high or low) fertility rates, which leads to (more or less) poverty?

6. Explain why, from 1875 to 2000, rapid population growth did not occur in developed countries but did occur in developing countries (see Fig. 7-7).

____

____

____

7. Which of the following social events prevented a population explosion in industrialized societies? (Check all that apply.)

[ ] reduction in infant and childhood mortality
[ ] availability of jobs in the cities
[ ] lower fertility rates
[ ] strict control of immigration

8. Fertility rates decline if development provides:

a. security in one's old age apart from ministrations of children (True or False)
b. lower infant mortality rates (True or False)
c. universal education for children (True or False)
d. opportunities for higher education and careers for women (True or False)

e. unrestricted access to contraceptives (True or False)

Development

9. What is the expected sequence of events from promoting development in low-income nations?

BETTER JOBS! BETTER ____________________ ! HIGHER STANDARD OF ____________________ ! EXPANDED ____________________ ! FOSTERING ____________________, ____________________, AND ____________________ ! LOWER ____________________ RATES.

Promoting the Development of Low-Income Countries

10. The World Bank has influenced the course of development in poor nations for the past __________ years.

Past Successes and Failures of the World Bank

11. Score the World Bank's successes [+] and failures [-] in promoting development in low-income nations using the following list.

[ ] increased GNP for some countries
[ ] social progress in some countries
[ ] declining fertility rates in some countries
[ ] achievement of replacement level fertility rates in some countries
[ ] the number of people living in absolute poverty
[ ] electric generating facility in India
[ ] large cattle operations in Latin America
[ ] mechanized plantations for growing cash crops

12. Indicate whether the following are [+] or are not [-] the working conditions in industrial plants in developing nations.

[ ] union-protected workforce
[ ] low wages and no fringe benefits
[ ] enforced pollution regulations
[ ] enforced child labor laws

13. Indicate whether the following are [+] or are not [-] benefits of World Bank loans applied to agriculture in developing countries.

[ ] higher production per acre
[ ] crop diversity
[ ] year-round food availability
[ ] direct farmer-to-consumer exchange
[ ] maintains the traditions of subsistence farming

The Debt Crisis

14. The national debt obligation of developing nations is (growing, shrinking).

15. What happens when a developing nation cannot even pay the interest on its loan?

____

16. What are three strategies used by developing nations to make even partial payment on their debt?

a. ____________________ b. ____________________

c. ____________________

17. How is the debt crisis in developing nations the essence of nonsustainability?

____

18. Give three examples of the large-scale centralized projects funded by the World Bank.

a. ____________________ b. ____________________

c. ____________________

19. (True or False) Large-scale centralized projects have alleviated absolute poverty conditions.

World Bank Reform

20. List two major differences in how the World Bank does business in developing countries today compared to the past.

a. ____

b. ____

A New Direction for Development - Social Modernization

21. Identify those factors that are are characteristics of the process of social modernization. (Check all that apply.)

[ ] good health and nutrition for mothers and children
[ ] old-age security apart from children
[ ] educational opportunities for women
[ ] availability of contraceptives
[ ] career opportunities
[ ] rights of women
[ ] improving resource management

Education

22. List two benefits of providing more education in developing countries.

a. ____

b. ____________________

Improving Health

23. Health care in developing countries means: (Check all that apply.)

[ ] bypass surgery
[ ] good nutrition and hygiene
[ ] proper treatment of infections
[ ] AIDS prevention and care

Family Planning

24. Indicate whether the following practices are [+] or are not [-] characteristic of family planning services and information.

[ ] counseling on reproduction, hazards of STDs, and contraceptive techniques
[ ] supplying contraceptives
[ ] counseling on pre- and post-natal health of mother and child
[ ] counseling on the health advantages of spacing children
[ ] promoting as many abortions as possible
[ ] nursing a baby for as long as possible
[ ] how to avoid unwanted or high-risk pregnancies

25. Which of the following elements are confounding the process of helping women in developing countries to lower their fertility? (Check all that apply.)

[ ] inadequate government funding
[ ] lack of access to and cost of contraceptives
[ ] inadequate sources of information on contraceptives

Enhancing Income

26. What three factors prevent most people in developing countries from obtaining a loan?

a. ____________________ b. ____________________

c. ____________________

27. What is the Grameen Bank doing to provide start-up money for people in developing countries?

____

28. Who are the preferred customers of the Grameen Bank? ____________________

29. Projects funded by the Grameen Bank:

a. do not upset the existing social structure (True or False)
b. involve a high degree of training or skill (True or False)
c. utilize local resources (True or False)

d. utilize centralized workplaces (True or False)
e. allow individuals to develop self-reliance (True or False)

Improving Resource Management

30. (True, False) People in developing countries lack basic skill in resource management.

Putting it All Together

31. What are the expected sequence of events from breaking the cycle of poverty, high fertility, and environmental degradation in low-income nations?

    BETTER HEALTH! ECONOMIC ____________________ ! BETTER ____________________ ! DELAYED ____________________ ! FEWER ____________________

The Cairo Conference

32. State the conceptual framework of all goals established in the 1994 ICPD Program of Action.

    ____

    ____

33. What part of the population (men, women, children) is particularly targeted in the 1994 ICPD Program of Action?

34. How will the Program of Action be funded?

    ____

NO VOCABULARY DRILL FOR THIS CHAPTER!

SELF-TEST

Indicate whether the following social events have happened in [a] developed, [b] developing, or [c] both countries.

1. [ ] reduction in infant and childhood mortality
2. [ ] availability of jobs in the cities
3. [ ] lower fertility rates

Indicate whether the following statements are true [+] or false [-] concerning the majority of projects funded by the World Bank.

4. [ ] relatively easy to administer, measure, monitor, and to demonstrate end products
5. [ ] result in substantial increases in the gross national product
6. [ ] result in substantial increases in wealth for all residents of the developing country
7. [ ] aggravate poverty
8. [ ] require the use of modern machinery and technology
9. [ ] increases diversity in crops and methods of farming
10. [ ] promote nonsustainable agriculture
11. [ ] increase the debt crisis in developing countries

12. Explain why developing nations are experiencing a population explosion and developed nations are not.

____

13. Give three reasons why U.S. firms are moving their production facilities to developing nations.

a. ____

b. ____

c. ____

14. Explain how the movement of U.S. firms to developing nations has the potential of doing more harm than good for the people.

____

____

Indicate whether the following are [+] or are not [-] characteristics of World Bank loans applied to agriculture in developing countries.

15. [ ] cash crops 16. [ ] higher production per acre 17. [ ] use of large tractors

18. [ ] combat absolute poverty

19. The utilization of local resources without upsetting the social structure is a characteristic of projects funded by the (World, Grameen) Bank.

20. List the three factors that are **most** directly correlated with low fertility rates.

a. ______________________________

b. ______________________________

c. ______________________________

21. Family planning services do not provide

a. counseling on reproduction and contraceptive techniques
b. contraceptives
c. unlimited abortions
d. counseling on the health advantages of spacing children

ANSWERS TO STUDY GUIDE QUESTIONS

1. population growth will exceed food-producing capacity of nations; 2. agricultural technology, declining fertility rates; 3. controlling population, economic development; 4. b, a, c, c, d, a, b; 5. more, high, more; 6. Developed countries experienced simultaneous declines in birth- and death rates, but developing countries had extended high birthrates with lowering death rates; 7. check first three, last one blank; 8. all true; 9. incomes; living; markets; trade, cooperation, peace; fertility; 10. 50; 11. -, +, +, -, -, -, -, -; 12. -, +, -, -; 13. all -; 14. growing; 15. unpaid interest is added to principal - shrink deeper into debt; 16. cash crops, austerity measures, exploitation of natural resources; 17. They liquidate capital assets to raise cash for short-term needs; 18. electric power plants, transportation - super highways, mechanized agriculture; 19. false; 20. environmental sensitivity, focus on the needs of the poor; 21. all checked; 22. empowers people to get more information, enhances economic strength of country; 23. check all but the first box; 24. +, +, +, +, -, +, +; 25. all checked; 26. credit risks, loans too small, status of women; 27. micro lending; 28. women; 29. true, false, true, false, true; 30. true; 31. productivity, education, marriage, children; 32. to create an economic, social, and cultural environment in which all people, regardless of age, race, or gender, can equitably share a state of well-being; 33. women and children; 34. 0.7 percent of developed nation's GNP

ANSWERS TO SELF-TEST

1. c; 2. a; 3. a; 4. +; 5. -; 6. -; 7. +; 8. +; 9. -; 10. +; 11. +; 12. The people of developing nations were unable to enhance their quality of life by moving to the city and finding jobs, which led to high fertility rates and a population explosion; 13. low wages, weak child labor laws, weak environmental laws; 14. exploits the poor by keeping them poor with low wages and social and environmental degradation; 15. +; 16. -; 17. +; 18. -; 19. Grameen; 20. availability of education for women, availability of contraceptives, good health and nutrition for mothers and children; 21. c

CHAPTER 8

SOIL AND THE SOIL ECOSYSTEM

---

1. Indicate the most probable means by which people in [a] developed and [b] developing nations obtain food.

   [ ] raise the food or gather it from natural ecosystems
   [ ] purchase the food
   [ ] have food given to them

2. List the five golden rules of sustainable agriculture.

   a. ____________________ b. ____________________ c. ____________________

   d. ____________________ e. ____________________

3. Conversion of farmland to nonfarm uses in the United States has averaged __________ acres/year over the past decade.

4. During the last 40 years, one third of Earth's cropland has been lost to __________________ and ____________________.

5. More than __________ percent of the world food supply is from land-based systems that rely on __________________ as the fundamental resource.

6. The cornerstone of sustaining civilization is maintaining productive __________________.

Plants and Soil

7. Productive topsoil involves the dynamic interactions among __________________, __________________, and __________________ (see Fig. 8-3).

Soil Characteristics

8. Match the soil horizons (A, B, C, E, O) with the soil profile characteristics listed below (see Fig. 8-4). Which horizon:

   [ ] contains the parent mineral material that originally occupied the site
   [ ] is high in iron, aluminum, calcium and other minerals
   [ ] contains the dead organic matter (detritus) deposited by plants
   [ ] is called topsoil
   [ ] is called subsoil
   [ ] stands for eluviation
   [ ] contains humus
   [ ] would be the first one lost due to water or wind erosion

9. Loam has a mineral composition of roughly _________ percent sand, __________ percent silt, and __________ percent clay.

10. As a soil becomes more coarse (average particle size increasing from clay to coarse sand), indicate whether each of the following soil properties will increase [+], decrease [-], or remain the same [0].

   [ ] infiltration
   [ ] surface runoff
   [ ] water-holding capacity
   [ ] aeration
   [ ] nutrient-holding capacity
   [ ] workability

11. Match the soil classes on the right with the characteristic or example on the left.

[ ] a tropical or subtropical rainforest soil	a. mollisols
[ ] soils of drylands and deserts	b. oxisols
[ ] soils of moist temperate forest biome	c. alfisols
[ ] soils of temperate grassland biomes	d. aridisols
[ ] have a deep A horizon rich in humus	
[ ] have a layer of iron and aluminum in the B horizon	

Soil and Plants

12. List the three basic ingredients of productive topsoil (see Fig. 8-3).

    a. ______________________ b. ______________________

    c. ______________________

13. What constitutes the base of the food chain in soil ecosystems? ______________

14. The soil environment must supply the plant or at least its roots with ______________, ______________, and ______________.

15. Three mineral nutrients required by the plant are ______________, ______________, and ______________.

16. The original source of mineral nutrients is from the breakdown of ______________ through the process of ______________.

17. (True or False) The process of weathering is fast enough to sustain vigorous plant growth without other sources.

18. In natural ecosystems, the greatest source of mineral nutrients for sustaining plant growth is from ______________.

19. (True or False) Plants need a more or less continuous supply of water to replace that lost in transpiration.

20. In addition to the frequency and amount of precipitation, the amount of water that is actually available to plants will depend on

    a. the amount of water that infiltrates versus running off (True or False)
    b. the amount of water that is held in the soil versus percolating through it (True or False)
    c. the amount of water that evaporates from the soil surface (True or False)

21. Indicate whether the following conditions need to be maximized [+] or minimized [-] in order to provide plants with the largest amount of available water.

    [ ] runoff [ ] infiltration [ ] water-holding capacity [ ] surface evaporation

22. Most plants have access to oxygen through the (roots, stems, leaves).

23. List two ways of depriving plants of sufficient oxygen.

    a. ______________________ b. ______________________

24. pH is a measure of relative ______________ and ______________.

25. Most plants require a soil environment that is

a. acidic b. basic or alkaline c. close to neutral

26. Neutral is expressed by a pH of __________.

27. Indicate the direction [a. in, b. out, c. equally in both directions] that water would go across the cell membrane given the following salt concentrations *outside* when compared to *inside* the plant cell wall.

a. [ ] higher b. [ ] lower c. [ ] same

Soil as an Ecosystem

28. List the five ingredients required of productive topsoil.

a. ____

b. ____

c. ____

d. ____

e. ____

29. The workability of clay soils will be (more or less) difficult when compared to sandy soils.

30. Distinguish between humus [H] and detritus [D] in the following definitions.

[ ] accumulation of dead leaves and roots on and in the soil
[ ] the residue of undigested organic matter that remains after dead plant material has been consumed

31. The first trophic level of the food web in soil ecosystems is

a. producers b. detritus c. humus d. decomposers

32. The second trophic level of the food web in soil ecosystems is

a. producers b. detritus c. humus d. decomposers

33. As soil organisms feed on the detritus and reduce it to humus, their activity also mixes and integrates humus with mineral particles developing what is called soil __________________.

34. The feeding and burrowing activities of organisms, which are feeding directly and indirectly on detritus,

a. break the detritus down to humus (True or False)
b. mix and integrate the humus with the mineral particles of the soil (True or False)

c. cause the soil to become loose and clumpy (True or False)
d. result in the formation of topsoil (True or False)

35. Indicate whether the presence of humus does [+] or does not [-] improve the following soil characteristics.

[ ] infiltration
[ ] water-holding capacity
[ ] nutrient-holding capacity
[ ] aeration
[ ] workability

36. Who gets what in the symbiotic relationship between some plants and mycorrhizae?

Plants: ______________________________

Mycorrhizae: ______________________________

37. (True or False) Green plants support the soil organisms by being the direct or indirect source of all their food.

38. (True or False) Soil organisms support green plants by making the soil more suitable for their growth.

39. As humus and the clumpy soil structure break down (i.e., mineralization), indicate whether the following increase [+], decrease [-], or remain the same [0].

[ ] infiltration
[ ] aeration
[ ] runoff
[ ] nutrient-holding capacity
[ ] water-holding capacity
[ ] leaching of nutrients
[ ] the ability of the soil to support plants

Soil Degradation

Erosion and Desertification

40. Erosion refers to soil particles being picked up by ______________________ or ______________________.

41. Loss of topsoil leads to

a. increased sediment deposits in rivers (True or False)
b. flooding in lowlands (True or False)
c. a continual cycle of erosion (True or False)

42. The erosion sequence [rank 1 to 3] from precipitation is:

[ ] gully [ ] splash [ ] sheet

43. Which of the following categories of soil particles would remain after wind or water erosion?

[ ] clay [ ] humus [ ] silt [ ] fine sand [ ] coarse sand and stones

44. Which of the following categories of soil particles are the most important soil components for nutrient and water-holding capacity?

[ ] clay [ ] humus [ ] silt [ ] fine sand [ ] coarse sand and stones

45. Uncontrolled wind and water erosion results in a soil ecosystem that is a functional ____________________.

46. (True or False) Once destroyed, topsoil will replenish itself.

Drylands

47. Dry land ecosystems cover more than _________ percent of the land area of Earth.

48. What is happening to over 70 percent the dry land ecosystems?

____

49. List two goals of the Convention to Combat Desertification (UNCCB).

a. ____

b. ____

Causing and Correcting Erosion

50. List the three major causes of soil erosion and desertification.

a. __________________ b. __________________ c. __________________

51. The basic reason for plowing and cultivation is for ____________________ control.

52. The basic drawback to plowing is that soil is exposed to __________________ and __________________ erosion.

53. (True or False) Plowing improves aeration and infiltration.

54. Indicate whether the following soil characteristics increase [+] or decrease [-] with continual applications of inorganic fertilizers.

[ ] organic matter

[ ] nutrient content
[ ] soil organisms
[ ] mineralization
[ ] desertification
[ ] nutrient-holding capacity
[ ] waterway pollution

55. (True or False) Chemical fertilizer is an adequate substitute for all the benefits provided by organic fertilizer.

56. Indicate whether the following statements are true [+] or false [-] concerning inorganic (chemical) fertilizers.

[ ] can be added to the soil efficiently and economically
[ ] are adequate substitutes for detritus
[ ] cause soil organisms to starve
[ ] enhance the mineralization process

57. Overgrazing leads to

a. less detritus to generate soil humus (True or False)
b. a gradual mineralization of the soil (True or False)
c. decreased grass production (True or False)
d. increased soil erosion (True or False)
e. decreased rainfall in the region (True or False)

58. Explain how BLM leases for grazing rights are a variation on the tragedy of the commons.

____

59. A forest cover

a. breaks the fall of raindrops reducing splash erosion (True or False)
b. allows water to infiltrate a litter-covered, loose topsoil (True or False)
c. allows 50 percent less runoff when compared to grasslands (True or False)
d. reduces leaching of soil nitrogen (True or False)

60. Which of the following is not a usual reason for clearing forests?

a. to experiment with secondary succession
b. to permit agriculture
c. to obtain structural wood
d. to obtain firewood

61. Deforestation causes MORE or LESS

a. __________ biodiversity
b. __________ erosion
c. __________ new seed germination
d. __________ sediment deposits in aquatic ecosystems

62. Tropical rain forests are being cleared at nearly __________ percent/year.

63. The total worldwide loss of soil from crop, range, and deforested lands is ___________ billion tons/year.

64. Indicate whether the following conditions increase [+] or decrease [-] from the deposition of sediments in waterways.

[ ] flooding
[ ] secondary succession
[ ] fish kills
[ ] pollution
[ ] groundwater reserves

Irrigation and Salinization

65. Adding water by artificial means is called __________________.

66. Irrigated water contains at least 200-500 parts per million salt, which, when left behind through evaporation leads to __________________.

67. Worldwide, an estimated __________ million acres is lost yearly to salinization and waterlogging.

68. In the United States, about _________ acres of original agricultural land has been rendered nonproductive by _____________________.

Addressing Soil Degradation

69. Making agriculture sustainable will depend upon our ability to control ________________, _________________, _________________, and __________________.

70. Identify the two levels of soil conservation.

a. ____________________ b. ___________________

Public Policy and Soils

71. Indicate whether the following descriptions of methods of controlling soil erosion apply to [a] contour farming, [b] strip cropping, [c] shelter belts, or [d] terracing (see Figure 8-19).

[ ] planting rows of trees around fields
[ ] plowing and cultivating at right angles to the slope
[ ] planting alternative strips of grass between strips of corn
[ ] grading slopes into a series of steps

72. List the four objectives of sustainable agriculture.

a. ____________________

b. ____________________

c. ____________________

d. ____________________

73. Indicate whether the following conservation measures were part of the [a] Conservation Reserve Program of 1985, [b] the Food Security Act of 1985, [c] 1988 Low Input Sustainable Agriculture (LISA) program, or [d] the Federal Agricultural Improvement and Reform Act of 1996.

[ ] farmers were paid to put highly erodible land into a Conservation Reserve program
[ ] farmers were required to develop and implement a soil conservation program
[ ] encourages farmers to use alternative farming methods
[ ] gave farmers more control in deciding what to plant with heavy reliance on market demands

74. Which of the following practices meet the objectives of sustainable agriculture? (Check all that apply.)

[ ] use of organic fertilizers
[ ] use of inorganic fertilizers exclusively
[ ] crop rotation
[ ] strip cropping
[ ] drip irrigation
[ ] dryland farming
[ ] use of chemical pesticides
[ ] biological control of pests
[ ] monoculture farming

75. Relate the following practices with [T] traditional or [A] alternative farming practices.

[ ] feed lots [ ] polycropping [ ] monocultures [ ] lower $ inputs

76. Past national farm policies promoted (traditional, alternative) farming practices.

77. Present national farm policies promote (traditional, alternative) farming practices.

VOCABULARY DRILL

Directions: Match each term with the list of definitions and examples given in the tables below. Term definitions and examples might be used more than once.

Term	Definition	Example(s)	Term	Definition	Example(s)
compaction			organic fertilizer		
composting			overcultivation		
deforestation			overgrazing		
desertification			pH		
erosion			salinization		
evaporative water			sediments		
fertilizer			sheet erosion		
gully erosion			soil aeration		
humus			soil fertility		
infiltrate			soil profile		
inorganic fertilizer			soil structure		
irrigation			soil texture		
loam			splash erosion		
mineral nutrients			subsoil		
mineralization			topsoil		
mycorrhizae			transpiration		
no-till agriculture			water-holding capacity		
nutrient-holding capacity			weathering		
			workability		

VOCABULARY DEFINITIONS
a. soil's ability to support plant growth
b. nutrients present in rocks
c. capacity to bind and hold nutrient ions
d. gradual chemical-physical breakdown of rock
e. material that contains one or more necessary nutrients
f. organic material that contains one or more necessary nutrients
g. inorganic material that contain one or more necessary nutrients
h. soil's ability to hold water after it infiltrates
i. transpiration: loss of water through plant leaves
j. artificial means of providing water to croplands
k. soil's ability to allow water to soak in
l. water loss from the soil surface
m. soil's ability to allow diffusion of oxygen and carbon dioxide
n. packing of soil
o. measure of hydrogen ion concentration in solution
p. relative proportions of sand, silt, and clay in soil
q. a specified mixture of sand, silt, and clay
r. workability: the ease with which soil can be cultivated
s. residue of partially decomposed organic matter
t. process of fostering decay of organic wastes under more-or-less controlled conditions
u. a loose, clumpy characteristic of soil
v. a dark humus-rich soil
w. a light brown, humus-poor soil
x. a layering of top and subsoil
y. soil organisms in a symbiotic relationship with plant roots
z. decomposition or oxidation of humus
aa. process of increasing soil loss through clear-cut logging
bb. effect of topsoil loss
cc. erosion when runoff water forms rivulets and streams
dd. process of reducing soil cultivation
ee. process of increasing soil loss through cultivation

VOCABULARY DEFINITIONS
ff. process of increasing soil loss through grazing too many livestock
gg. intolerable increase in soil salinity
hh. soil particles released from topsoil during erosion
ii.the process of soil and humus being picked up and carried away by wind or water
jj. water erosion resulting from lack of infiltration
kk. impact of falling raindrops

VOCABULARY EXAMPLES	
a. optimum amounts of mineral nutrients, water, oxygen	l. Fig. 8-4
b. phosphate, potassium, calcium	m. fungi
c. mulch	n. Fig. 8-13
d. Fig. 8-7	o. clear-cut logging
e. Fig. 8-21	p. Fig. 8-15
f. oxygen concentration	q. Fig. 8-14 , splash and sheet not shown
g. excessive foot or vehicular traffic	r. Fig. 8-18
h. 0 ↔ 7 ↔ 14	s. plowing, discing, harrowing
i. 40 : 40 : 20 sand, silt, clay, respectively	t. BLM lands
j. Fig. 8-9	u. Fig. 8-22
k. making humus	v. muddy river

SELF-TEST

1. Which trophic level would contain detritus, decomposers, earthworms, soil protozoans, and ants?

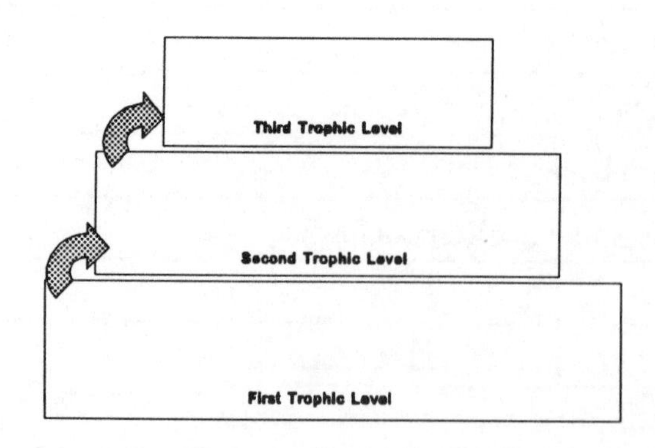

Complete the table below by indicating the relationship of the factors listed with the soil properties given in the table. The relationship can be designated as excellent [E], good [G], medium [M], or poor [P].

FACTORS

Soil Property	Water Infiltration	Water-holding Capacity	Nutrient-holding Capacity	Aeration
Sand	2. _____	3. _____	4. _____	5. _____
Silt	6. _____	7. _____	8. _____	9. _____
Clay	10. _____	11. _____	12. _____	13. _____
Loam	14. _____	15. _____	16. _____	17. _____
Humus rich	18. _____	19. _____	20. _____	21. _____

Explain how the practices of inefficient wood burning and scouring the landscape of leaves, sticks, manure, and other forms of detritus have a negative effect on the structure and function of the soil ecosystem.

22. Negative effect on STRUCTURE: ____

____

23. Negative effect on FUNCTION: ____

____

24. Which of the agricultural practices below will increase [+] or decrease [-] soil erosion?

[ ] plowing
[ ] using inorganic fertilizers exclusively
[ ] overgrazing
[ ] deforestation
[ ] planting shelter belts
[ ] strip cropping

Complete the table below first by identifying those agricultural practices that do [+] and do not [-] meet the objectives of sustainable agriculture and second by indicating whether these practices are traditional [T] or alternative [A] types.

Agricultural Practices	Sustainable	Type
Use of organic fertilizers	25. [ ]	26. [ ]
Center pivot irrigation	27. [ ]	28. [ ]

Use of chemical pesticides	29. [ ]	30. [ ]
Dryland farming	31. [ ]	32. [ ]
Crop rotation	33. [ ]	34. [ ]

ANSWERS TO STUDY GUIDE QUESTIONS

1. a, a, b; 2. keeping the soil covered, using minimum or zero tillage, using mulch to provide plant nutrients, maximizing biomass production, maximizing biodiversity; 3. 1.4 million; 4. erosion, salinization; 5. 90, soil; 6. soil; 7. soil organisms, detritus, and mineral particles; 8. C, B, O, A, B, E, O, O; 9. 40, 40, 20; 10. +, -, -, +, -, +; 11. b, d, c, a, a, b; 12. soil organisms, detritus, mineral particles; 13. detritus; 14. mineral nutrients, water, oxygen; 15. phosphate, potassium, calcium; 16. rocks, weathering; 17. false; 18. detritus; 19. true; 20. all true; 21. -, +, +, -; 22. roots; 23. overwatering, compaction; 24. acidity, alkalinity; 25. c; 26. 7; 27. b, a, c; 28. good supply of nutrients and nutrient-holding capacity, good infiltration and water-holding capacity, porous structure for aeration, ph = 7, low salt concentration; 29. more; 30. D, H; 31. b; 32. d; 33. structure; 34. all true; 35. all +; 36. plants get needed nutrients from fungal decomposition of detritus, fungi get some nutrients from plant roots; 37. true; 38. true; 39. -, -, +, -, -, +, -; 40. wind, water; 41. all true; 42. 3, 1, 2; 43. coarse sand and stones; 44. clay and humus; 45. desert; 46. true, but only if given enough time and no disturbance; 47. 33; 48. being affected by erosion and desertification; 49. fund projects to reverse land degradation, disseminate traditional knowledge on effective dryland agricultural practices; 50. overcultivation, overgrazing, deforestation; 51. weed; 52. wind, water; 53. false; 54. -, +, -, +, +, -, +; 55. false; 56. +, -, +, +; 57. all true; 58. too many cattle on grassland because grazing lease cost so low; 59. all true; 60. a; 61. less, more, less, more; 62. 1; 63. 23; 64. +, -, +, +, -; 65. irrigation; 66. salinization; 67. 10; 68. 400,000, salinization; 69. erosion, overgrazing, deforestation, salinization; 70. individual landowner, public policy; 71. c, a, b, d; 72. reduce use of chemicals, keep food safe and wholesome, maintain a productive topsoil, keep agriculture economically viable; 73. a, b, c, a, d; 74. check, blank, check, check, check, check, blank, check, blank; 75. T, A, T, A; 76. traditional; 77. alternative

ANSWERS TO SELF-TEST

1. see Figure 8-8; 2 to 17. see Table 8-2; 18 to 21. all excellent; 22. the detritus structure is lost which forms the base of the soil ecosystem food chain and one of the primary soil erosion preventatives; 23. without detritus for detritus feeders and decomposers, the transfer of energy and nutrients to other soil biota cannot take place; 24. +, +, +, +, -, -; 25. +; 26. A; 27. -; 28. T; 29. -; 30. T; 31. +; 32. A; 33. +; 34. A

VOCABULARY DRILL ANSWERS					
Term	**Definition**	**Example(s)**	**Term**	**Definition**	**Example(s)**
compaction	n	g	organic fertilizer	f	c, j
composting	t	k	over cultivation	ee	s
deforestation	aa	o	overgrazing	ff	t
desertification	bb	p	pH	o	h
erosion	ii	q	salinization	gg	u
evaporative water	l	d	sediments	hh	v
fertilizer	e	c	sheet erosion	jj	q
gully erosion	cc	q	soil aeration	m	f
humus	s	j	soil fertility	a	a
Infiltrate	k	d	soil profile	x	l
inorganic fertilizer	g	c	soil structure	u	l
irrigation	j	e	soil texture	p	I
loam	q	i	splash erosion	kk	q
mineral nutrients	b	b	subsoil	w	l
mineralization	z	n	topsoil	v	l
Mycorrhizae	y	m	transpiration	i	d
no-till agriculture	dd	r	water-holding capacity	h	d
nutrient-holding capacity	c	c	weathering	d	c
			workability	r	i

CHAPTER 9

WATER: HYDROLOGIC CYCLE AND HUMAN USE

Water - a Vital Resource

1. Water covers _________ percent of Earth's surface.

2. __________ percent of Earth's water is saltwater.

3. Two-thirds of the freshwater is bound up in:

   a. ____________________ and b. ____________________

4. Only __________ percent is accessible freshwater.

5. Water supply shortages loom in at least _________ countries.

The Hydrologic Cycle

6. Earth's water cycle is also called the _______________ cycle.

7. Water enters the atmosphere through _______________ and _______________.

8. Water returns from the atmosphere through _______________ and _______________.

Evaporation, Condensation, and Purification

9. The three physical states of water are ___________________, ___________________, and ___________________.

10. These three states are the result of different degrees of _________________ bonding between water molecules.

11. What are the two forces that determine the different degrees of interaction between water molecules?

    a. ______________________ b. ______________________

12. Hydrogen bonding tends to (attract, separate) water molecules.

13. Kinetic energy tends to (attract, separate) water molecules.

14. Below freezing, (hydrogen bonding, kinetic energy) is the dominant force determining the association between water molecules.

15. Above freezing, (hydrogen bonding, kinetic energy) is the dominant force determining the association between water molecules.

16. During evaporation, kinetic energy is (high, low) when compared to hydrogen bonding.

17. During condensation, kinetic energy is (high, low) when compared to hydrogen bonding.

18. Evaporation and condensation are two water purification processes similar to ____________________.

19. (True or False) All the natural freshwater on Earth comes from a distillation process.

20. Water vapor enters the atmosphere by way of:

   a. ____________________ from all water and moist surfaces.

   b. ____________________ which is evaporation through the leaves of plants.

21. The amount of water vapor held in the air at various temperatures is called ____________________ humidity.

22. Warm air holds (more or less) water than cold air.

Precipitation

23. Rising air results in (high, low) precipitation whereas descending air results in (high, low) precipitation.

24. Air rises over equatorial regions and descends over subequatorial regions (see Fig. 9-5a and 5b), therefore:

   a. rainfall is (high, low) in equatorial regions
   b. rainfall is (high, low) in subequatorial regions

25. The rainshadow (see Fig. 9-6) refers to a region of (high, low) precipitation.

Water Over and Through the Ground

26. When precipitation hits the ground, what two alternative pathways might it follow?

   a. ____________________ b. ____________________

27. All ponds, lakes, streams, and rivers are referred to as ____________________ waters.

28. Water that infiltrates may follow either of two pathways, including:

   a. ____________________ or b. ____________________ water.

29. Plants draw mainly from ____________________ water.

30. Water that percolates down through the soil eventually comes to an ____________________ layer and accumulates, filling all the empty pores and spaces. This accumulated water is now called ____________________ water, and its upper surface is called the ____________________ table.

31. The layers of porous material through which groundwater moves are called an ________________________.

32. The point where water actually enters an aquifer is called the ________________ area.

33. Natural exits of groundwater to the surface are ________________ or ______________.

34. Springs and seeps feed streams and rivers, becoming part of ________________ water.

Summary of the Water Cycle

35. Summarize the events in the water cycle by matching the processes or stages listed on the left with the letters (a-j) in the water cycle diagram. Some letters may be used more than once individually or in a sequence (e.g., h, i, j).

[ ] evaporation
[ ] condensation
[ ] transpiration
[ ] percolation
[ ] surface runoff
[ ] cloud formation
[ ] precipitation
[ ] capillary water
[ ] gravitational water
[ ] aquifer
[ ] groundwater
[ ] infiltration
[ ] surface runoff loop
[ ] groundwater loop
[ ] evapotranspiration loop

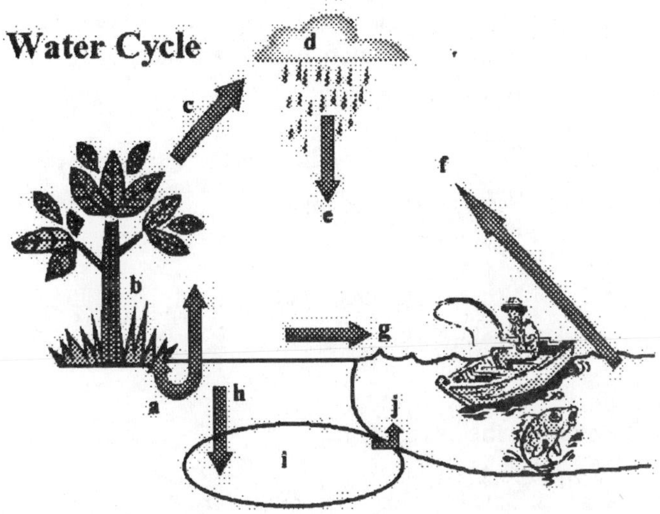

36. The two filters for water that infiltrates through the ground are ________________ and ________________.

37. Match the terms on the left with the water descriptions on the right (see Table 9-1).

Water Terms	Water Descriptions
[ ] fresh	a. a mixture of fresh and salt water
[ ] salt	b. salt concentration < 0.01 percent
[ ] brackish	c. salt concentration at least 3 percent
[ ] hard	d. high mineral content
[ ] soft	e. low mineral content

Human Impacts on the Water Cycle

38. List the three main impacts that humans have on the water cycle.

a. ____

b. ____

c. ____

Changing the Surface of Earth

39. List three ways humans have changed the surface of Earth leading to alterations in the hydrological cycle.

    a. ____

    b. ____

    c. ____

40. Deforestation leads to (more or less) infiltration, (more or less) groundwater recharge, (more or less) runoff, (more or less) sedimentation, (more or less) water pollution, and (more or less) flooding.

41. Destruction of wetlands leads to (more or less) infiltration, (more or less) groundwater recharge, (more or less) runoff, (more or less) sedimentation, (more or less) water pollution, and (more or less) flooding.

42. Urbanization leads to (more or less) infiltration, (more or less) groundwater recharge, (more or less) runoff, (more or less) sedimentation, (more or less) water pollution, and (more or less) flooding.

Polluting the Water Cycle

43. How have humans polluted the three major filter loops in the hydrological cycle?.

Pollutants	Filter loop
Acid rain	a. Evapotranspiration Loop
Road salt	b. Surface runoff Loop
Landfills	c. Groundwater Loop
Deforestation	

Sources and Uses of Freshwater

44. The two primary human concerns regarding freshwater are ________________ and ______________ water.

45. What proportion (percent) of Earth's water resources are used for:

    a. irrigation _________ b. industry _________ c. human consumption __________

46. The major source of freshwater is (a. groundwater b. surface water).

47. Center pivot irrigation systems use huge amounts of (surface, ground) water and may use as much as (1, 10, 100) thousand gallons per minute.

Overdrawing Water Resources

Consequences of Overdrawing Surface Waters

48. No more than __________ percent of a river's average flow can be taken risking shortfalls.

49. However, the demand on some rivers exceeds _________ percent of the average flow.

50. The consequences of overdrawing surface waters include:

    a. wetlands along and at the mouth of the river drying up (True or False)
    b. estuaries at the mouth of the river becoming increasingly salty (True or False)
    c. dieoffs of large populations of wildlife (True or False)

Consequences of Overdrawing Groundwater

51. Depletion of groundwater will very likely result in

    a. cutbacks in agricultural production (True or False)
    b. diminishing surface water (True or False)
    c. adverse effects on stream and river ecosystems (True or False)
    d. land subsidence (True or False)
    e. formation of sinkholes (True or False)
    f. saltwater intrusion into wells in coastal regions (True or False)
    g. required cutbacks in municipal consumption (True or False)

Obtaining More Water

52. It may be possible to increase water supplies:

    a. at modest cost (True, False)
    b. without severe ecological impacts (True, False)

53. Future water shortages will

    a. become increasingly common (True or False)
    b. be of longer duration (True or False)
    c. affect increasingly large areas and numbers of people (True or False)
    d. go away (True or False)

Using Less Water

54. The greatest demands on U.S. freshwater (consumptive, nonconsumptive) uses are particularly in (see Table 9-2 & Fig. 9-11):

    a. agriculture   b. electrical power production   c. industry   d. homes

Irrigation

55. About _________ percent of irrigation water is wasted through evaporation.

56. A water-conserving alternative to surface and sprinkling irrigation is _______________ irrigation.

Municipal Systems

57. Water consumption in modern homes averages around _________ gallons per person per day.

58. The most effective conservation technique would be to use (more, the same, less) water.

59. An example of recycling water is ________________ water.

Desalting Sea Water

60. What are the two technologies used for desalinization:

    a. ________________________________ b. ________________________________

61. What is the cost per 1,000 gallons of desalinized water? _________

62. Farmers currently pay $________ per 1,000 gallons of freshwater for irrigation.

Storm Water

63. As land is developed, natural soil surfaces are replaced by various hard surfaces. By this action

    a. infiltration is (increased or decreased).
    b. runoff is (increased or decreased).

Mismanagement and Its Consequences

64. List three consequences of stormwater mismanagement.

    a. ___________________ b. ___________________ c. ___________________

65. As a result of decreases in infiltration and increases in runoff due to urbanization, predict whether each of the following will increase [+], decrease [-], or remain the same [0].

    [ ] amount of groundwater recharge
    [ ] level of water table
    [ ] amount of flow from springs
    [ ] stream flow between rains
    [ ] stream flow during rains
    [ ] stream bank erosion
    [ ] sediment deposition in the stream channel
    [ ] breadth of stream channel
    [ ] depth of stream channel
    [ ] pollution of stream channel
    [ ] frequency of flooding
    [ ] heights of floods

66. List the categories of pollutants that enter streams and rivers from surface runoff.

a. ______________________ b. ____________________ c. ____________________

d. ______________________ e. ____________________ f. ____________________

g. ______________________

67. To alleviate the problem of urban flooding , streams may be channelized. In this process

a. a channel is dug out (True or False)
b. curves are smoothed out (True or False)
c. the channel is lined with concrete or rock. (True or False)
d. a natural water flow is restored (True or False)
e. a natural stream ecology is restored (True or False)

Improving Storm Water Management

68 Describe one technique for improving storm water management (see Fig. 9-25).

____

Water Stewardship

69. Current trends in water use (are, are not) sustainable.

70. List four stakeholders in future water use decisions.

a. ____________________________ b. ____________________________

c. ____________________________ d. ____________________________

VOCABULARY DRILL

Directions: Match each term with the list of definitions and examples given in the tables below. Term definitions and examples might be used more than once.

Term	Definition	Example(s)	Term	Definition	Example(s)
aquifer			polluted water		
capillary water			precipitation		
channelization			rainshadow		
condensation			recharge area		
consumptive water use			relative humidity		
desalinization			saltwater		
drip irrigation			saltwater intrusion		
evaporation			sinkhole		
evapotranspiration			spring		
freshwater			storm water management		
gravitational water			surface runoff		
gray water			surface waters		
groundwater			transpiration		
humidity			water cycle		
hydrological cycle			water purification		
infiltration			water quality		
land subsidence			water quantity		
nonconsumptive water use			water table		
percolation			water vapor		
physical states of water			watershed		
			xeroscaping		

VOCABULARY DEFINITIONS
a. amount of freshwater
b. purity of freshwater
c. water molecules entering the atmosphere
d. water conditions given different temperatures
e. water molecules rejoining by hydrogen bonding
f. salt content less than 100 ppm
g. water molecules in the air
h. amount of water vapor in the air
i. amount of water as a percentage of what air can hold at that temperature
j. salt content more than 100 ppm
k. containing elements or particulate matter that is beyond the norm
l. process of removing unwanted elements or particulate matter
m. dry region downwind of a mountain
n. all the land area that contributes water to a particular stream or rive
o. nonenclosed water impoundments
p. water held in the soil
q. water that soaks into the ground
r. water rising to the atmosphere through evaporation or transpiration and leaving it through condensation and precipitation
s. water movement in the gaseous state promoted by Earth's heat or through plants
t. the effect of condensation and cooling temperatures
u. movement of excess noninfiltrated water
v. accumulation of water in an impervious layer of rock or dense clay
w. water subjected to the pull of gravity
x. movement through the soil beyond capillary water
y. the upper surface of accumulated groundwater
z. where water enters an aquifer
aa. layers of porous material through which groundwater moves
bb. where water exits the ground as a significant flow from a relatively small opening
cc. lost for further human use
dd. results of a collapsed underground cavern drained of it supporting groundwater

VOCABULARY DEFINITIONS
ee. gradual settling of land when the water table drops
ff. movement of saltwater into a freshwater aquifer
gg. remains available for further human use
hh. a system of irrigation that applies small steady amount of water to the plant roots
ii. water that has been used for other purposes and reused again
jj. change in the structural characteristics of stream beds
kk. landscaping with desert species that require no additional watering
ll. converting salt into freshwater
mm. movement of water through plants
nn. procedures used to offset the destructive effects of stormwater runoff

VOCABULARY EXAMPLES	
a. 2.5 percent	p. rivulets, streams
b. nonpolluted and < 100ppm salt	q. Ogallala Aquifer
c. water in gaseous state	r. infiltrated water not available to plant roots
d. gas, liquid, solid	s. western Nebraska grasslands
e. water going from gaseous to liquid state	t. irrigation
f. water vapor + temperature	u. Fig. 9-17
g. oceans, seas	v. Mono Lake
h. acid rain	w. Fig. 9-18
i. desalinization	x. gray water
j. Death Valley	y. Fig. 9-19
k. Missouri River Basin	z. washing machine, shower water
l. lakes, rivers	aa. Fig. 9-24
m. water available to plant roots	bb. cacti, mesquite bush
n. Fig. 9-3	cc. Middle East - Saudi Arabia
o. rain, snow	dd. Fig. 9-25

SELF-TEST

1-15. Illustrate the hydrological cycle, including all filtration loops and relevant terms. For example, give yourself one point for each correct representation of: surface runoff loop, evaporation-transpiration loop, groundwater loop, water vapor, condensation, cloud formation, transpiration, precipitation, evaporation, infiltration, percolation, water table, groundwater, springs, surface runoff.

16. Indicate whether the depletion of groundwater will [+] or will not [-] lead to:

[ ] increased groundwater use in agriculture.
[ ] subsidence.
[ ] saltwater intrusion.
[ ] increased municipal consumption.
[ ] decreased flow through seeps and springs.
[ ] adverse effects on stream and river ecosystems.

17. As land is developed, natural soil surfaces are replaced by various hard surfaces. As a result of this change indicate whether the following will increase [+], decrease [-], or remain the same [0].

[ ] amount of groundwater recharge
[ ] level of water table
[ ] amount of spring flow

[ ] stream flow during rains
[ ] frequency of flooding
[ ] pollution of stream channel

Explain how human activities have adversely affected the following filtration loops in the hydrological cycle.

18. Surface runoff loop: ____________________

19. Evapotranspiration loop: ____________________

20. Groundwater loop: ____________________

ANSWERS TO STUDY GUIDE QUESTIONS

1. 71; 2. 97; 3. polar ice caps, glaciers; 4. 0.77; 5. 26; 6. hydrological; 7. evaporation, transpiration; 8. condensation, precipitation; 9. gas, liquid, solid; 10. hydrogen; 11. kinetic energy, hydrogen bonding; 12. attract; 13. separate; 14. hydrogen bonding; 15. kinetic energy; 16. high; 17. low; 18. distillation; 19. true; 20. evaporation, transpiration; 21. relative; 22. more; 23. high, low; 24. high, low; 25. low; 26. infiltration, runoff; 27. surface; 28. capillary, gravitational; 29. capillary; 30. impervious, groundwater; 31. aquifer; 32. recharge; 33. springs, seeps; 34. surface; 35. f; d; c; h; g; d; e; a; h; i; i; a; e, g, f, d; e, a, b, c, d ; e, h, i, j, f, d; 36. soil, porous rock; 37. b, c, a, d, e; 38. changing Earth's surface, pollution, overwithdrawals; 39. deforestation, draining of wetlands, urban sprawl; 40. less, less, more, more, more, more; 41. same as #40; 42. same as #40; 43. a, b, c, a; 44. supply, purity; 45. 70, 20, 10; 46. a; 47. ground, 10; 48. 30; 49. 100; 50. all true; 51. all true; 52. all false; 53. true, true, true, false; 54. nonconsumptive, c; 55. 30-50; 56. drip; 57. 106; 58. less; 59. gray; 60. microfiltration/reverse osmosis, distillation; 61. $3; 62. $0.02; 63. decreased, increased; 64. flooding, streambank erosion, increased pollution; 65. -, -, -, -, +, +, +, +, -, +, +, +; 66. fertilizer, "cides," bacteria, salt, toxic chemicals, oil and grease, litter; 67. true, true, true, false, false; 68. stormwater retention reservoirs; 69. are not; 70. natural ecosystem, industry, agriculture, domestic

---

ANSWERS TO SELF-TEST

1-15. (see Figure 9-3); 16. -, +, +, -, +, +; 17. -, -, -, +, +, +; 18. Large quantities and different types of pollutants are part of urban runoff; 19. Air pollutants become mixed with water molecules during condensation and precipitation (e.g., acid rain); 20. Depletion of groundwater and pollutants in soil that are carried into groundwater

VOCABULARY DRILL ANSWERS					
Term	**Definition**	**Example(s)**	**Term**	**Definition**	**Example(s)**
aquifer	aa	n, q	polluted water	k	h
capillary water	p	m	precipitation	t	n, o
channelization	jj	aa	rainshadow	m	j
condensation	e	e	recharge area	z	s
consumptive water use	cc	t	relative humidity	i	f
desalinization	ll	cc	saltwater	j	g, v
drip irrigation	hh	y	saltwater intrusion	ff	w
evaporation	c	c, n	sinkhole	dd	u
evapotranspiration	s	n	spring	bb	n
freshwater	f	b	stormwater management	nn	dd
gravitational water	w	n, r	surface runoff	v	n, p
gray water	ii	z	surface waters	o	l
groundwater	v	n, q	transpiration	mm	n
humidity	h	c	water cycle	r	n
hydrological cycle	r	n	water purification	l	i
infiltration	q	n	water quality	b	b
land subsidence	ee	u	water quantity	a	a
nonconsumptive water use	gg	x	water table	y	n, q
percolation	x	n	water vapor	g	c
physical states of water	d	d	watershed	n	k
			xeroscaping	kk	bb

CHAPTER 10

THE PRODUCTION AND DISTRIBUTION OF FOOD

Crops and Animals: Major Patterns of Food Production

The Development of Modern Industrialized Agriculture

1. Indicate whether the following practices were [+] or were not [-] common agricultural practices in the United States 150 years ago.

   [ ] bringing additional land into cultivation
   [ ] use of inorganic fertilizers
   [ ] groundwater irrigation
   [ ] use of chemical pesticides
   [ ] substituting new genetic varieties of grains
   [ ] rotating crops
   [ ] growing many different kinds of crops
   [ ] recycling animal wastes

The Transformation of Traditional Agriculture

2. What other revolution contributed to a revolution in agriculture? _________________

3. Today, less than __________ percent of the U.S. workforce produces enough food for the nation's needs.

4. Indicate whether the following practices were [+] or were not [-] characteristics of the agricultural revolution that increased food production so dramatically from 1950 to 1990.

   [ ] bringing additional land into cultivation
   [ ] increasing use of fertilizer
   [ ] increasing use of irrigation
   [ ] increasing use of chemical pesticides
   [ ] substituting new genetic varieties of grains
   [ ] rotating crops
   [ ] growing many different kinds of crops
   [ ] recycling animal wastes

5. Which of the following agriculture practices will [+] and will not [-] increase food production in the year 2000 and beyond?

   [ ] bringing additional land into cultivation
   [ ] increasing use of fertilizer
   [ ] increasing use of irrigation
   [ ] increasing use of chemical pesticides
   [ ] substituting new genetic varieties of grains
   [ ] rotating crops
   [ ] growing many different kinds of crops
   [ ] recycling animal wastes

6. The shift from animal labor to machinery has led to:

a. high dependency on fossil fuel energy (True or False)
b. soil compaction (True or False)
c. recycling of animal wastes (True or False)

7. Any future attempts to bring additional land under cultivation will probably result in:

a. use of marginal, highly erodible lands (True or False)
b. loss of forests and wetlands (True or False)

8. Adding more fertilizer than the optimum required leads to:

a. ever-increasing productivity (True or False)
b. increased plant vulnerability to pests (True or False)
c. pollution (True or False)

9. The use of pesticides in agriculture:

a. provided better pest control (True or False)
b. increased crop yields (True or False)
c. led to pest resistance (True or False)
d. caused adverse side effects to human and environmental health (True or False)

10. Even though irrigated acreage increased about 2.6 times from 1950 to 1980, irrigation is increasing at a slower pace due to:

a. lack of new water sources (True or False)
b. development of drought-resistant crops (True or False)
c. waterlogging and accumulation of salts in the soil (True or False)

11. The new varieties of wheat and rice produced in the 1960s required more ________________ and ________________ but gave yields ________________ to ________________ when compared to traditional varieties.

12. The high worldwide production of wheat and rice resulting from the new genetic varieties was hailed as the ________________ Revolution.

13. Identify which of the following aspects of the Green Revolution were benefits [B] or drawbacks [D] for the people of developing nations.

[ ] closed the gap between food production and food needs in some countries
[ ] the growth and production limits of "high-yielding" varieties of wheat and rice
[ ] heavy reliance on irrigation
[ ] heavy reliance on fertilizers
[ ] impact on farm laborers and small landowners
[ ] impact on culturally - specific crops

Subsistence Agriculture in the Developing World

14. Using the terms MORE or LESS, complete the following statements about subsistence farming. Subsistence farming in developing countries:

is ________________ labor-intensive, requires ________________ technology, utilizes ________________ marginally productive land, and involves the clearing of ________________ tropical forests.

Animal Farming and Its Consequences

15. What proportion of the world's crop lands is used to feed animals? ___________

16. In the United States, ___________ percent of the grain crop goes to animals.

17. Indicate whether the following are [+] or are not [-] environmental consequences of animal farming.

[ ] overgrazing
[ ] deforestation of tropical rain forests
[ ] increases of atmospheric carbon dioxide

Prospects for Increasing Food Production

18. List two prospects for increasing food production in the future.

a. ____

b. ____

19. List two factors that will limit the ability to increase crop yields.

a. ______________________ b. ______________________

The Promise and Problems of Biotechnology

20. Explain genetic engineering in terms of three results.

a. ____

b. ____

c. ____

21. The products of genetic engineering:

a. will be affordable to farmers in developing countries (True or False)
b. are considered safe by all consumers (True or False)
c. will be part of agriculture as we know it in the twenty-first century (True or False)
d. will keep food production in pace with population growth (True or False)

Food Distribution and Trade

Patterns in Food Trade

22. The world region recognized as the major source of exportable grains is ____________________ ____________________.

23. Identify whether the 1998 grain import (I) export (E) ratio of each of the following countries (see Table 10-3) is [a] I = E, [b] I < E, or [c] I > E.

    [ ] North America [ ] Asia [ ] Latin America [ ] Africa

24. For those countries that import more grain than they export, what has been the primary cause for the trade imbalance?

    ____

Levels of Responsibility in Supplying Food

25. Identify the three major levels of responsibility for meeting food needs.

    a. ____________________ b. ____________________ c. ____________________

26. Indicate whether the following policies focus on meeting the food needs of [a] family, [b] nation, or [c] globe (see Fig. 10-9).

    [ ] effective safety net
    [ ] food aid for famine relief
    [ ] employment security
    [ ] effective family planning
    [ ] fair trade
    [ ] just land distribution

Hunger, Malnutrition, and Famine

Nutrition vs. Hunger

27. The major nutritional problems in children are ____________________ and people of all ages are ____________________

Extent and Consequences of Hunger

28. At least __________ of the people on Earth suffer from the effects of hunger and malnutrition.

29. In what order (1 to 3) are the consequences of malnutrition and hunger reflected in a population?

    __________ men __________ children __________ women

Root Cause of Hunger

30. The root cause of hunger is _________________.

31. (True or False) Our planet can produce enough food for everyone alive today.

32. Food surpluses enter the cash economy and flow in the direction of ________________ demand, not of _________________ need.

33. Using the answers to question 32, is it MORE or LESS probable that:

    __________ pet cats will be well fed
    __________ undernourished children will be well fed

Famine

34. The two primary causes of famine are _________________ and _______________.

35. Indicate whether the primary famine in the following countries was due to [a] drought, [b] war, [c] both, or [d] government incompetence.

    [ ] Sahel region of West Africa
    [ ] Mozambique
    [ ] Sudan
    [ ] North Korea

36. Explain FEWS in terms of:

    What it means: ____

    What it does: ____

Food Aid

37. Indicate whether the following statements are true [+] or false [-] concerning the overall effects of providing food aid to countries in need. Providing food:

    [ ] actually alleviates chronic hunger in developing countries
    [ ] undercuts prices for domestically produced food sold in local markets
    [ ] causes the local economy to deteriorate
    [ ] contributes to environmental and ecological deterioration

38. Food aid to developing countries usually (disrupts, stimulates) the local economy and leads to a/an (increase, decrease) in local food production.

39. Food aid tends to (improve, aggravate) the problem of chronic hunger.

Final Thoughts on Hunger

40. (True or False) The agricultural practices of the past 40 years are sustainable.

41. In order to alleviate world hunger, humankind needs to develop new (a. technologies, b. forums of political and social action).

42. In order to alleviate world hunger, humankind needs to distribute food based on (a. need, b. ability to pay).

VOCABULARY DRILL

Directions: Match each term with the list of definitions and examples given in the tables below. Term definitions and examples might be used more than once.

Term	Definition	Example(s)	Term	Definition	Example(s)
absolute poverty			high-yielding		
cash crops			hunger		
exports			imports		
famine			malnutrition		
food aid			organic farming		
food security			subsistence farming		
green manures			sustainable agriculture		
Green Revolution			undernutrition		

VOCABULARY DEFINITIONS
a. lack of sufficient income to meet most basic biological needs
b. crops sold by a country
c. products sold by a country
d. severe shortage of food accompanied by significant increase in death rate
e. distribution of food where famine is evident
f. ability to meet the nutritional needs of all family members
g. a remarkable increase in crop production
h. crop residue
i. crop's ability to produce beyond the norm through genetic manipulation
j. lack of basic food required for energy and meeting nutritional needs
k. products purchased by a country
l. lack of essential nutrients

VOCABULARY DEFINITIONS
m. regular addition of crop residues and animal manures to build up soil humus
n. farming for household needs and small cash crops
o. maintain agricultural production without degrading the environment
p. lack of adequate food energy (calories)

VOCABULARY EXAMPLES	
a. 1.2 billion people	g. lawn clippings
b. coffee, cocaine, cocoa	h. health foods
c. Fig. 10-11	i. Fig. 10-5
d. Egypt	j. applying the four principles of ecosystem sustainability
e. the family level of responsibility	k. 824 million people
f. Fig. 10-4b	

SELF-TEST

Which of the following factors will enable humankind to increase agricultural production in the next 40 years? (Check all that apply.)

1. [ ] bringing additional land into cultivation

2. [ ] increasing use of fertilizer

3. [ ] increasing use of irrigation

4. [ ] increasing use of chemical pesticides

5. [ ] substituting new genetic varieties of grains

Briefly explain your responses to items 1 through 5 above.

6. Bringing additional land into cultivation:____

____

7. Increasing use of fertilizer: ____

____

8. Increasing use of irrigation:____

____

9. Increasing use of chemical pesticides: ____

____

10. Substituting new genetic varieties of grains: ______________________________________

____

11. Identify [+] those characteristics of agricultural that will be absolutely necessary to initiate future Green Revolutions in developing countries.

[ ] reliance on "high-yielding" varieties of wheat and rice

[ ] polycropping techniques

[ ] application of the principles of ecosystem sustainability

[ ] more mobilization of the human workforce in food production

[ ] utilization of culturally - specific crops

[ ] reliance on irrigation and fertilizers

12. There appears to be a direct relationship between increased reliance on food imports and ________________ growth in some developing countries.

Formulation of policies to meet the food needs of the global population are required at the levels of:

13. __________________ 14. ____________________ 15. __________________

16. Explain how the global priorities on food distribution make it more probable that a pet cat rather than a starving child will be well fed.

____

____

17 Explain, from the agribusiness side of the issue, why food aid to developing countries does not solve but rather aggravates their hunger problems.

____

____

ANSWERS TO STUDY GUIDE QUESTIONS

1. -, -, -, -, -, +, +, +; 2. industrial; 3. 3; 4. +, +, +, +, +, -, -, -; 5. -, -, -, -,-, +, +, +; 6. true, true, false; 7. true, true; 8. false, true, true; 9. all true; 10. true, false, true; 11. fertilizer, water, double, triple; 12. green; 13. B, D, D, D, D, D; 14. more, less, more, more; 15. 25%; 16. 70; 17. all +; 18. continue to increase crop yields, grow more food instead of cash crops; 19. loss of cropland through soil erosion, climate changes; 20. crossbreeds of genetically different plants, incorporation of desired traits into crop lines and animals, cloning of domestic animals; 21. false, false, true, true; 22. North America; 23. b, c, c, c; 24. population growth surpassed food production; 25. family, nation, globe; 26. b, c, a, a, c, b; 27. lack of proteins and some vitamins, calories for food energy; 28. half; 29. 3, 1, 2; 30. poverty; 31. true; 32. economic, nutritional; 33. more, less; 34. war, drought; 35. a, b, b, c; 36. famine early warning system, measures trends in vegetation and rainfall in Africa in order to predict famines; 37. -, +, +, +; 38. disrupts, decreases; 39. aggravate; 40. false; 41. b; 42. a

---

ANSWERS TO SELF-TEST

1 - 5. all blank, cannot rely on any of these methods to increase agricultural production in the next 40 years; 6.Tthere is not additional tillable land left to bring into agriculture; 7. The optimum level is already being used, using more will not increase production; 8. Ground table water reserves are nearly depleted, no other large supplies of water are left; 9.Pest resistance capabilities and adverse effects on humans and the environment preclude continued use of pesticides; 10. There are no new laboratory-designed "high-yielding" varieties to use; 11. blank, +, +, +, +, blank; 12. population; 13. family; 14. nation; 15. global; 16. Surpluses are distributed based on economic priorities rather than the nutritional needs of people, i.e., more money can be made from turning surpluses into cat food than feeding starving people; 17. Free food aid undercuts the peasant farmer's ability and willingness to grow crops and free food is not distributed equitably or to those who need it most

VOCABULARY DRILL ANSWERS					
Term	**Definition**	**Example(s)**	**Term**	**Definition**	**Example(s)**
absolute poverty	a	a	high-yielding	i	f
cash crops	b	b	hunger	j	c
exports	c	b	imports	k	b
famine	d	c	malnutrition	l	c
food aid	e	d	organic farming	m	h
food security	f	e	sustainable agriculture	o	j
green manures	h	g	subsistence farming	n	i
Green Revolution	g	f	undernutrition	p	k

CHAPTER 11

WILD SPECIES: BIODIVERSITY AND PROTECTION

Value of Wild Species

1. What is the estimated worth of the goods and services provided by wild species?
   __________ trillion

Biological Wealth

2. We presently know of the existence of _________ million species of plants, animals, and microbes.

3. How many plant and animal species have become extinct in the United States? ___________

Two Kinds of Values

4. ____________________ value exists if an organism's existence or use benefits some other entity.

5. Intrinsic ____________ value considers the organism's value for its own sake.

6. List the four values of natural species.

   a. ____

   b. ____

   c. ____

   d. ____

Sources for Agriculture, Forestry, Aquaculture, and Animal Husbandry

7. Indicate whether the following statements pertain to [a] wild populations of plants or animals or [b] cultivated (cultivar) populations of plants or animals.

   [ ] highly adaptable to changing environmental conditions
   [ ] highly nonadaptable to changing environmental conditions
   [ ] have numerous traits for resistance to parasites
   [ ] lack genetic vigor
   [ ] can only survive under highly controlled environmental conditions
   [ ] have a high degree of genetic diversity in the gene pool of the population
   [ ] have a low degree of genetic diversity in the gene pool of the population
   [ ] represents a reservoir of genetic material commonly called a genetic bank

8. Of the estimated 7,000 plant species existing in nature, how many have been used in agriculture?

   ____

9. What is the significance of the winged bean (Fig. 11-2) in regard to your answer to question 8?

   ____

Sources for Medicine

10. Of what value is the chemical called vincristine?

    ____

11. Vincristine is an extract from the ____.

12. Of what value is the chemical called capoten?

    ____

13. Capoten is an extract from the ____.

14. (True or False) There is the potential of discovering innumerable drugs in natural plants, animals, and microbes.

Recreational, Aesthetic, and Scientific Value

15. One of the largest income-generating enterprises in many developing countries is ___________________.

16. As leisure time (increases, decreases), larger portions of the economy become connected to supporting activities related to the natural environment.

17. Pollution (increases, decreases) the commercial benefits of the natural environment.

18. The contribution of wildlife-related recreation to the U.S. economy was calculated to be more than $ __________ billion in 1996.

19. Indicate whether the following are [a] recreational, [b] aesthetic, [c] scientific or [d] intrinsic values of the biota of natural ecosystems.

[ ] hunting
[ ] sport fishing
[ ] hiking
[ ] camping
[ ] being in a forest rather than a city dump
[ ] study of ecology
[ ] just knowing that certain plant and animal species are alive

Intrinsic Value

20. Intrinsic value arguments center on ______________ rights and ______________.

Saving Wild Species

Game Animals in the United States

21. (Commercial or Regulated) hunting caused the extinction or near extinction of some game animals.

22. Restrictions on hunting that represent wildlife management are

a. requiring a hunting license (True or False)
b. limiting the type of weapon used in the harvest (True or False)
c. hunting within specified seasons (True or False)
d. setting size and/or sex limits on harvested animals (True or False)
e. limiting the number that can be harvested (True or False)
f. prohibiting commercial hunting (True or False)

23. Which of the following do [+] and do not [-] represent contemporary wildlife management problems?

[ ] near extinction of the wild turkey
[ ] road-killed animals
[ ] urban wildlife, e.g., opossums, skunks, raccoons
[ ] lack of natural predators
[ ] wildlife as vectors for certain diseases
[ ] cougar predation on humans
[ ] urban Canada geese populations
[ ] pets predation by coyotes

The Endangered Species Act

24. Snowy egrets were heavily exploited in the 1800s for their (a. meat, b. eggs, c. nests, d. feathers) that were used in (a. fancy restaurants, b. ladies' hats, c. curiosity shops, d. grocery stores).

25. The two states that were first to pass laws protecting plumed birds were ____________________ and ______________________.

26. The law that forbids interstate commerce in illegally killed wildlife is called the __________________ Act.

27. The law that protects species from extinction is called the ____________________ __________________ Act of ______________________.

28. Indicate whether the following provisions are strengths [+] or weaknesses [-] of the Endangered Species Act.

[ ] the need for official recognition
[ ] stiff fines for the killing, trapping, uprooting, or commerce of endangered species
[ ] controls on government development projects in critical habitats
[ ] enforcement of the act
[ ] recovery programs
[ ] taxonomic status of some animals protected by the act

29. Identify two endangered bird species that were recovered.

a. ____________________________ b. ____________________________

30. Explain the connection between sandhill and whooping cranes.

____

31. What are two major concerns of opponents of the ESA that surfaced in the spotted owl controversy?

a. ____________________________ b. __________________________

Biodiversity

32. Over geological time, the net balance between _______________ and _______________ has favored the gradual (accumulation, loss) of species.

33. The number of plant and animal species on Earth today is estimated at __________ million, with the greatest species richness found in _______________ plants and _______________.

34. Which country has at least 5 percent of all living species?

    ____

The Decline of Biodiversity

35. How many species of animals _______________ and plants _______________ have become extinct since the early days of colonization?

36. Globally, how many species of animals _______________ and plants _______________ are in danger of becoming extinct?

37. In what type of ecosystem can one find the greatest biodiversity? _______________

38. List four sources of evidence that a decline of biodiversity is occurring.

    a. ______________________________ b. ______________________________

    c. ______________________________ d. ______________________________

39. Give two reasons why species on oceanic islands are more susceptible to extinction than continental species.

    a. ____

    b. ____

Reasons for the Decline

40. List five factors that represent evidence of or contribute to the decline in biodiversity.

    a. ____

    b. ____

    c. ____

    d. ____

    e. ____

41. Match the following examples of factors that represent evidence of or contribute to the decline of biodiversity with those you listed [a through e] in the previous question.

Examples	Factors
fragmentation	a. habitat alterations
clear-cut logging	b. human population growth
expansion of the human population over the globe	c. pollution
the brown tree snake	d. exotic introductions
the greenhouse effect	e. overuse
prospects of huge immediate profits	
the growing fad for exotic pets, fish, reptiles, birds, and house plants	
simplification	
endocrine disrupters	
black bear gallbladders	

42. Indicate whether the following examples of habitat alteration represent [a] conversion, [b] fragmentation, or [c] simplification.

[ ] stream channelization
[ ] shopping malls
[ ] highway development

43. Africa and Asia have lost more than __________ of their original natural habitat.

44. As human population increases, species survival worldwide will (increase, decrease).

45. The greatest catastrophe to hit natural biota in 65 million years will be

a. acid rain b. human population growth c. global warming d. urbanization

46. Species adapt very (quickly, slowly) to environmental change.

47. Global warming is predicted to occur very (quickly, slowly).

48. Exotic species are responsible for _________ percent of all animal extinctions since 1600.

49. Usually, the introduction of exotic species into a country (increases, decreases) biodiversity.

50. Usually, an exotic species is a (more, less) effective competitor for resources.

51. Give three examples of exotic species introduced into the United States.

a. ________________ b. ________________ c. ________________

52. As certain plants and animals become increasingly rare, the price people are willing to pay will (increase, decrease).

53. Exploiters of natural biota will (encourage, discourage) public education on poaching and black market trade.

54. Below is a chain of events that perpetuates overuse of natural biota. Which link in the chain needs to be removed to most effectively curtail overuse?

a. hunter b. trading post c. taxidermist d. tourist shop e. consumer tourist

55. Give an example of human use of the following wildlife or plant species or their parts.

a. small songbirds: ______________________________

b. tropical hardwoods: ______________________________

c. bear gallbladders:_ ______________________________

d. parrots: ______________________________

Consequences of Losing Biodiversity

56. Explain why keystone species are such an important group to consider as a consequence of losing biodiversity.

____

International Steps to Protect Biodiversity

57. List two international steps that have been taken to protect biodiversity.

a. ______________________________ b. ______________________________

58. Which of the above steps

a. focuses on trade in wildlife and wildlife parts? ______________________________

b. focuses on conserving biological diversity worldwide? ______________________________

c. does not yet have political support in the United States? ______________________________

Stewardship Concerns

59. List four steps that must be taken to protect our biological wealth.

a. ______________________________

b. ______________________________

c. ______________________________

d. ______________________________

VOCABULARY DRILL

Directions: Match each term with the list of definitions and examples given in the tables below. Term definitions and examples might be used more than once.

Term	Definition	Example(s)	Term	Definition	Example(s)
anthropocentric			fragmentation		
biodiversity			genetic bank		
biological wealth			instrumental value		
biota			intrinsic value		
conversion			keystone species		
cultivar			simplification		
ecotourism			speciation		
endangered species			threatened species		
exotic species			vigor		
extinction					

VOCABULARY DEFINITIONS
a. reduced to a point at which it is in imminent danger of extinction
b. judged to be in jeopardy but not on brink of extinction
c. assemblage of all living organisms
d. assemblage of all living organisms and their ecosystems
e. existence benefits some other entity
f. an entity that derives instrumental value from biota
g. something that has value for its own sake
h. tolerance to adverse conditions
i. cultivated variety of a wild strain
j. source of all genetic traits of wild strains of plants and animals
k. visiting a place in order to observe unique ecological sites
l. the creation of new species
m. the disappearance of species
n. human development of natural areas
o. process of cutting natural areas into separated parcels
p. diminishing abiotic and biotic habitat diversity

VOCABULARY DEFINITIONS
q. species introduced into an area from somewhere else
r. a species whose role is absolutely vital for the survival of many other species in the ecosystem

VOCABULARY EXAMPLES	
a. 1.75 million species	i. housing subdivisions
b. channelization	j. humans
c. competitiveness	k. Table 11-2
d. corn, wheat, rice	l. Fig. 11-8
e. Darwin's finches	m. large predators
f. Fig. 11-1	n. cats
g. Galapagos Islands	o. sources for agriculture
h. highways	p. just knowing that it exists

SELF-TEST

1. Indicate whether the following species have basically [a] instrumental or [b] intrinsic value.

   [ ] tuna
   [ ] butterflies
   [ ] rosy periwinkle
   [ ] hummingbirds

2. Match the examples on the left with the values of natural species on the right.

Examples	Values
[ ] cultivars	a. intrinsic
[ ] captone	b. commercial
[ ] bird watching	c. recreational
[ ] ecotourism	d. medicinal
[ ] just knowing the species exists	e. agricultural

3. The intrinsic value of wild species is rooted in ________________ rights and ______________ beliefs.

4. Biodiversity is measured by the ________________ of different species in an ecosystem.

5. Which of the following types of ecosystems has the greatest biodiversity?

a. grassland b. tropical rain forest c. urban development d. cornfield

6. Indicate whether the following activities conserve [+] or destroy [-] biodiversity.

[ ] conversion of habitat
[ ] pollution
[ ] introduction of exotics
[ ] expansion of human population globally
[ ] overuse
[ ] regulated hunting
[ ] recovery plans
[ ] ecosystem management

Complete statements 7 and 8 using the terms MORE or LESS.

7. More expansion of the human population leads to __________ pollution, __________ use of wildlife species, and __________ biodiversity.

8. More introduction of exotics leads to __________ competition for resources, __________ predation on endemic prey species, and __________ biodiversity.

9. The consequence(s) of species extinctions is (are):

a. cascading loss of other species
b. loss of nature's services
c. loss of genetic banks
d. all of the above

10. It was (commercial, regulated) hunting that caused the extinction or near extinction of some wild animals.

Indicate whether the following statements pertain to [a] ESA, [b] CITES, [c] Convention on Biological Diversity or [d] more than one of these.

11. [ ] focuses on the trade of wildlife or their parts
12. [ ] uses recovery plans to re-establish species numbers
13. [ ] emphasizes sovereignty of a nation over its biodiversity

ANSWERS TO STUDY GUIDE QUESTIONS

1. 33; 2. 1.75; 3. 500; 4. instrumental; 5. intrinsic; 6. agriculture, medicine, aesthetics and recreation, intrinsic; 7. a, b, a, b, b, a, b, a; 8. 30; 9. a wild species that can become an alternative human food source; 10. treatment for leukemia; 11. rosy periwinkle; 12. controls high blood pressure; 13. Brazilian pit viper; 14. true; 15. ecotourism; 16. increases; 17. decreases; 18. 101; 19. a, a, a, a, b, c, d; 20. animal, religion; 21. commercial; 22. all true; 23. -, all the rest +; 24. d, b; 25. Florida, Texas; 26. Lacey; 27. Endangered Species, 1973; 28. -, +, +, -, -, -; 29. wild turkey, bald eagle; 30. Sandhill cranes are surrogate parents for whooping crane chicks; 31. property rights and jobs; 32. speciation, extinction, accumulation; 33. 1.75, flowering, insects; 34. Costa Rica; 35. 484, 654; 36. 5,400, 26,000; 37. tropical rain forest; 38. commercial fishing down, decline in waterfowl populations, disappearance of songbirds from some regions, overall decline in songbird populations; 39. small area limits population size, human intrusions most severe; 40. habitat alterations, human population growth, pollution, exotic introductions, overuse; 41.a, e, b, d, e, e, d, a, c, e; 42. c, a, b; 43. two-thirds; 44. decrease; 45. c; 46. slowly; 47. quickly; 48. 39; 49. decreases; 50. more; 51. house mouse, Norway rat, wild boar, donkey, horse, nutria, red fox, armadillo, pheasant, starling, house sparrow, honeybee, fire ant; 52. increase; 53. discourage; 54. e; 55. food, furniture, perceived medicinal value, exotic pets; 56. a species whose role is absolutely vital for the survival of many other species in an ecosystem; 57. CITES, Convention on Biological Diversity; 58. CITES, Convention, Convention; 59. reform policies that lead to declines in biodiversity, address the needs of people whose livelihood is derived from exploiting wild species, practice conservation at the landscape level, promote more research on biodiversity

---

ANSWERS TO SELF-TEST

1. a, b, a, b, e, d, c, b, a; 3. animal, religious; 4. number; 5. b; 6. -, -, -, -, -, +, +, +; 7. more, more, less; 8. more, more, less; 9. d; 10. commercial; 11. b; 12. a; 13. c

VOCABULARY DRILL ANSWERS

Term	Definition	Example(s)	Term	Definition	Example(s)
anthropocentric	f	j	fragmentation	o	h
biodiversity	d	a	genetic bank	j	a
biological wealth	d	a	instrumental value	c	o
biota	c	a	intrinsic value	g	p
conversion	n	i	keystone species	r	m
cultivar	i	d	simplification	p	b
ecotourism	k	g	speciation	l	e
endangered species	a	l	threatened species	b	k
exotic species	q	n	vigor	h	c
extinction	m	f			

CHAPTER 12

ECOSYSTEMS AS RESOURCES

Biological Systems in a Global Perspective

Major Systems and Their Services

1. Identify the largest (percent of total, Table 12-1) and smallest terrestrial ecosystems.

   Largest = ________________ Smallest = ________________

2. List seven services performed by natural ecosystems.

   a. ____

   b. ____

   c. ____

d. ____________________

e. ____________________

f. ____________________

g. ____________________

3. The loss of which natural service led to:

   a. siltation and flooding of rivers in Bangladesh: ____________
   b. eutrophication of Chesapeake Bay: ____________

4. How much would it cost to duplicate the water purification and fish propagation capacity of a single acre of tidal wetland? $ ____________ per year

5. (True or False) We can compensate for the losses of natural services.

Ecosystems and Natural Resources

6. In what two ways does society assign value to natural resources?

   a. ____________ b. ____________

7. The term natural resource is often defined in an (ecological, economic) setting.

Conservation and Preservation

8. Any resource that has the capacity to renew or replenish itself is called a ____________ resource.

9. (True or False) The term conservation implies complete denial of any use of a natural resource.

10. The aim of conservation is to ____________ or ____________ use.

11. (True or False) The term preservation implies complete denial of any use of a natural resource.

Patterns of Use of Natural Ecosystems

12. People harvesting natural resources in order to provide for their needs is called ____________ use.

13. The exploitation of ecosystem resources for economic gain is called ____________ use.

14. Match the following definitions with [a] maximum sustained yield and [b] carrying capacity.

    [ ] the maximum population that an ecosystem can support
    [ ] the maximum use a system can sustain without impairing its ability to renew itself

15. Maximum sustained yield is obtained with (optimal, maximum) population size.

16. Maximum sustained yield is (reached, exceeded) when use begins to destroy regenerative capacity.

17. Indicate whether the following population characteristics will increase [+] or decrease [-] sustained yield.

[ ] a population below or within the carrying capacity
[ ] a population that is approaching or exceeding carrying capacity

18. (True or False) Carrying capacity and optimal population size are constant.

19. Indicate whether the following procedures will increase [+] or decrease [-] sustained yield.

[ ] pollution and other forms of habitat alteration
[ ] use levels that ignore population characteristics in relation to carrying capacity
[ ] using suitable management procedures
[ ] harvesting populations that are at or exceeding the carrying capacity

20. Indicate whether the following procedures will sustain [+] or diminish [-] maximum sustained yield.

[ ] protecting biota from the "tragedy of the commons"
[ ] invoking regulations but without enforcement
[ ] reducing the economic incentives that promote violation of regulations
[ ] preserving habitats
[ ] protecting habitats from pollution

21. Which of the following actions or statements are [+] or are not [-] characteristic of the tragedy of the commons?

[ ] Whoever grazes the most cattle gains an economic advantage over those who graze less.
[ ] "If I don't harvest this resource, someone else will."
[ ] reacting to diminished catches of clams by harvesting less
[ ] whenever two or more independent groups are engaged in exploitation of a resource
[ ] Eventually no one will be grazing cattle or digging clams.

22. The tragedy of the commons can be avoided by limiting ________________.

23. What are two strategies for limiting access to the commons?

a. ______________________ b. ______________________

24. List the conditions for regulated access to the commons.

a. ____

b. ____

c. ____

Biomes and Ecosystems Under Pressure

25. List four ecosystems that are under pressure.

a. ______________________________ b. ______________________________

c. ______________________________ d. ______________________________

Forest Biomes

26. The major threat to the world's forests is total ____________________.

27. Worldwide, (a. 2/3, b. 1/4, c. 1/2) of the area originally covered by forests is now devoid of trees.

28. Indicate whether clearing forests increases [+] or decreases [-]:

[ ] productivity
[ ] nutrient recycling
[ ] biodiversity
[ ] soil erosion
[ ] transpiration

29. List the five principles of sustainable forestry management.

a. ______________________________

b. ______________________________

c. ______________________________

d. ______________________________

e. ______________________________

30. Between 1960 and 1990, ____________ percent of tropical rain forests were converted to other uses.

31. Tropical rain forests provide:

a. millions of plant and animal species (True or False)
b. a system that maintains the climate of Earth (True or False)
c. a system that controls air pollution (True or False)

32. List the two basic causes of deforestation.

a. ______________________________ b. ______________________________

33. Indicate whether the following conditions promote [+] or prevent [-] the destruction of the tropical rain forest.

[ ] colonization of forested lands
[ ] huge national debts
[ ] fast food chains and cheap hamburger
[ ] using forest for ecotourism rather than logs
[ ] extractive reserves
[ ] giving control of forests to indigenous villagers
[ ] Statement on Forest Principles

Ocean Ecosystems

34. Open oceans were traditionally considered an ____________________ commons.

35. As a response to over fishing in the international commons, many nations extended their _________________ limits.

36. In 1977 the United States extended its territorial limits from 3 to 12 miles to _________________ miles offshore.

37. Oceans could yield at least __________ million metric tons of food on a sustained basis.

38. What is the best strategy to restore fisheries like George's and Grand Bank?

____

39. List three basic problems common to all fisheries.

a. ____________________ b. ____________________ c. ____________________

40. What is the purpose of the 1996 Magnuson Act?

____

41. Indicate whether the following groups or events have had a [+] or [-] effect on whale populations.

[ ] International Whaling Commission
[ ] International Union for Conservation of Nature
[ ] whale watching
[ ] Japan's harvest of whales for scientific research
[ ] Stellwagen Bank

42. What three countries want whaling to continue?

a. ______________________________ b. ______________________________

c. ______________________________

43. Why do the above three countries want whaling to continue?

____

44. Today, (pollution, over harvest) is the greatest threat to freshwater and estuarine fisheries.

45. Indicate whether the following sources of damage impact coral reefs [C] or mangroves [M].

[ ] lucrative tropical fish trade
[ ] shrimp aquaculture
[ ] islander poverty
[ ] logging
[ ] warm water temperatures

46. Indicate whether the following statements represent values of coral reefs [C] or mangroves [M] or both [B].

[ ] protect coasts from storm damage and erosion
[ ] form a rich refuge and nursery for many marine fish
[ ] important food source for local people

Public and Private Lands in the United States

47. Nearly _________ percent of U.S. land is publicly owned.

48. The majority of U.S. public lands are in the _______________ states and _______________ (see Fig. 12-17).

49. The purpose of the Wilderness Act of 1964 is to (conserve, preserve)?

50. What is the goal of the Wilderness Act of 1964?

____

National Parks and National Wildlife Refuges

51. Identify the federal agencies that manage:

a. National Parks: ______________________

b. National Wildlife Refuges: ______________________

c. National Forests: ______________________ and ______________________

52. What is the purpose of the Greater Yellowstone Coalition (see Fig. 12-18)?

____

National Forests

53. Three-fourths of the managed commercial forests are (private, state, federal) lands.

54. (True or False) Deforestation is no longer a problem in the United States.

55. Only _________ percent of the original U.S. forests are left.

56. Most U.S. forests are (old, second) growth forests.

57. (True or False) U.S. taxpayers make money from the sale of national forest lumber.

58. List four factors that characterize "New Forestry".

    a. ____

    b. ____

    c. ____

    d. ____

59. What are the five principles of ecosystem management adopted by the Forest Service?

    a. ____

    b. ____

    c. ____

    d. ____

    e. ____

Private Land Trusts

60. List two ways private land trusts protect natural areas from development.

    a. ____

    b. ____

61. List three examples of private land trusts.

a. ____

b. ____

c. ____

62. The purpose of the private land trust is to

a. obtain development rights on private land (True or False)
b. purchase land to protect it from development (True or False)
c. obtain land at the expense of the landowners (True or False)
d. use the power of eminent domain to obtain land (True or False)

63. How many acres are protected by the Nature Conservancy? __________ acres

VOCABULARY DRILL

Directions: Match each term with the list of definitions and examples given in the tables below. Term definitions and examples might be used more than once.

Term	Definition	Example(s)	Term	Definition	Example s)
carrying capacity			natural resources		
commons			natural services		
conservation			preservation		
consumptive use			private land trust		
environmental backlash			renewable resource		
extractive reserves			restoration		
Fishery			tragedy of the commons		
maximum sustained yield					

VOCABULARY DEFINITIONS
a. land that is protected for native people
b. a limited marine area or a group of fish species being exploited
c. services provided by ecosystems
d. natural ecosystems and the biota in them
e. the harvesting of natural resources to meet people needs
f. capacity to replenish itself through reproduction
g. manage or regulate use so capacity of species to renew itself is not exceeded
h. ensure ecosystem continuity regardless of potential utility
I. resource owned by many people in common or by no one
j. exploitation of the commons leading to a loss of the resource
k. highest possible rate of use that is matched with rate of replacement
l. maximum population the ecosystem can support on a sustainable basis
m. repair damaged ecosystems to normal functions as representative flora and fauna
n. people and organizations that oppose environmental action
o. a nonprofit organization that will accept outright gifts of land or easements

VOCABULARY EXAMPLES	
a. 7.5 million acres of rain forest	f. Fig. 12-6a and b
b. Georges Bank, MA	g. Baytown, Texas
c. Fig. 12-3	h. National Wetlands Coalition
d. Fig. 12-4	i. Nature Conservancy
e. Muriquis monkey	

SELF-TEST

Match the following descriptions with one of the services provided by nature.

	Descriptions	Services
1.	preventing eutrophication	____________________
2.	providing plant cover and litter	____________________
3.	providing natural predators	____________________
4.	nutrient storage	____________________

5. Which of the following actions or statements is not characteristic of the tragedy of the commons?

   a. Whoever harvests the most gains an economic advantage over those who harvest less.
   b. "If I don't get it, someone else will".
   c. occurs whenever several independent groups are engaged in exploitation of a resource
   d. reduced harvests lead to reduced attempts to access the resource

6. It is (conservation, preservation) that implies complete denial of any use of the natural resource.

7. Maximum sustained yield occurs at the optimal population level. What is the optimum population level?

   ____

   ____

8. Forests are cleared for (ecological or economic) reasons.

9. Indicate whether the following are long [L] or short-term [S] consequences of clearing forests.

   [ ] loss of biodiversity
   [ ] soil erosion
   [ ] loss of nutrient recycling
   [ ] profits from housing developments

10. Which of the following conditions prevent the destruction of tropical rain forests?

   a. accumulation of huge national debts in Third World countries
   b. colonization of rainforest habitats by peasant farmers
   c. development of extractive reserves
   d. development of pasture lands

11. Explain why the ocean fishery was considered a commons.

   ____

   ____

12. In the last decade, consumptive use of whales has (increased, decreased) and nonconsumptive use has (increased, decreased).

13. Identify one nonconsumptive use of whales. ______________________

14. Explain how a federal redefinition of "wetland" decreased the amount of protected acreage.

   ____

   ____

15. Match the organizations with the ecosystems under their care.

[ ] land trusts — a. National Marine Fisheries Service
[ ] National Wildlife Refuges — b. Nature Conservancy
[ ] ocean fishery — c. U.S. Fish and Wildlife Service

List four patterns of human uses of natural resources.

16. ______________________ 17. ______________________

18. ______________________ 19. ______________________

ANSWERS TO STUDY GUIDE QUESTIONS

1. tundra and desert land, wetlands; 2. maintenance of hydrological cycle, modification of climate, absorption of pollutants, transformation of toxic chemicals, erosion control and soil building, pest management, maintenance of oxygen and nitrogen cycles, carbon storage and maintenance of carbon cycle; 3. erosion control, absorption of pollutants; 4. $100,000; 5. false; 6. economic, ecological; 7. economic; 8. renewable; 9. false; 10. manage, regulate use; 11. true; 12. consumptive; 13. productive; 14. b, a; 15. optimal; 16. exceeded; 17. +, -; 18. false; 19. -, -, +, +; 20. +, -, +, +, +; 21. +, +, -, +, +; 22. access; 23. private ownership, regulated access; 24. sustained benefits, fairness in access rights, common consent of the regulated; 25. forests and woodlands, tropical rain forests, oceans, coral reefs and mangroves; 26. removal; 27. c; 28. all -; 29. sustainable forestry, responsible practices, forest health and productivity, protecting special sites, continuous improvement; 30. 20; 31. all true; 32. economic development, human population growth; 33. +, +, +, +, -, -, -; 34. international; 35. territorial; 36. 200; 37. 117; 38. close fishery until it has time to restore itself; 39. too many boats, high technology, too few fish; 40. requires that depleted fisheries be rebuilt and that fishing be maintained at biologically sustainable levels; 41. +, +, +, -, +; 42. Japan, Norway, Iceland; 43. claim whale meat is an integral part of their diet; 44. pollution; 45. C, M, C, M, C; 46. all B; 47. 40; 48. western, Alaska; 49. preserve; 50. provides for permanent protection of undeveloped and unexploited areas so that natural ecological processes can operate freely; 51. National Park Service, U.S. Fish and Wildlife Service, National Forest Service, Bureau of Land Management; 52. conserve the larger ecosystem that surrounds Yellowstone National Park; 53. Private; 54. true; 55. 5; 56. second; 57. false; 58. cutting trees less frequently, leaving wider buffer zones along streams, leaving dead logs and debris, protecting broad landscapes; 59. takes integrated view of terrestrial and aquatic ecosystems, integrates variety of spatial scales, incorporates landscape ecology, allows for change in management, incorporates the human element; 60. accept land as outright gifts, purchase land; 61.Land Trust Alliance, Nature Conservancy, Trustees of Reservations in Massachusetts; 62. true, true, false, false; 63. 71 million

---

ANSWERS TO SELF-TEST

1. absorption of pollutants; 2. erosion control and soil building; 3. pest control; 4. carbon storage; 5. d; 6. preservation; 7. a population level above that which would reduce yield by decreased population and below that which would reduce yield by resource competition; 8. economic; 9. L, L, L, S; 10. c; 11. it is a resource owned by many people in common or no one; 12. decreased, increased; 13. whale watching; 14. the new definition required more (21) consecutive days of water inundation which removed 30 million wetland acres that fit the original definition (7 consecutive days); 15. b, c, a; 16. tragedy of the commons; 17. conservation or preservation; 18. restoration; 19. maximum sustained yield.

VOCABULARY DRILL ANSWERS					
Term	**Definition**	**Example(s)**	**Term**	**Definition**	**Example(s)**
carrying capacity	l	f	natural resources	d	b
commons	i	b	natural services	c	c
conservation	g	e	preservation	h	e
consumptive use	e	d	private land trust	o	i
environmental backlash	n	h	renewable resource	f	b
extractive reserves	a	a	restoration	m	g
fishery	b	b	tragedy of the commons	j	b
maximum sustained yield	k	f			

CHAPTER 13

ENERGY FROM FOSSIL FUELS

Energy Sources and Uses

Harnessing Energy Sources: An Overview

1. Throughout human history, the major energy source was _________________ labor.

2. The primary limiting factor to machinery designs in the early 1700s was a _________________ source.

3. The _________________ engine launched the Industrial Revolution in the late 1700s.

4. The first major fuel for steam engines was (a. coal, b. wood, c. gasoline, d. electricity).

5. By the 1800s the major fuel for steam engines was (a. coal, b. wood, c. gasoline, d. electricity).

6. By 1920, coal provided __________ percent of all energy used in the United States.

7. Three 1800 technologies that provided an alternative to coal were the ________________ engine, ________________ drilling technology, and the ability to refine ________________ oil.

8. Crude oil provides ____________ percent of the total U.S. energy demand.

9. Coal provides _________ percent of the total U.S. energy demand.

10. Indicate whether the following attributes are [A] advantages or [D] disadvantages of oil-based fuels when compared to coal.

    [ ] portability [ ] pollution [ ] energy content [ ] known reserves

Electrical Power Production

11. Electricity is a (primary, secondary) energy source.

12. One primary energy source for generating electrical power is (a. coal b. wood c. oil-based fuels).

13. Indicate whether the following are advantages [A] or disadvantages [D] of electrical power.

    [ ] pollution from secondary energy source
    [ ] pollution from primary energy sources
    [ ] habitat alterations
    [ ] environmental effects of mining and processing
    [ ] efficiency of secondary energy source

14. Thermal production of electricity is only _________ percent efficient.

15. __________ percent of the primary energy is lost in the form of ______________ causing ________________ pollution in aquatic ecosystems.

16. Explain how the following forms of generating electricity represent a transference of pollution from one place to another.

    a. coal-burning power plants: ____

    b. hydroelectric power plants: ____

    c. nuclear power plants: ____

Matching Sources to Uses

17. Identify the four major categories of energy use (see Fig. 13-9).

    a. ______________________________ b. ______________________________

    c. ______________________________ d. ______________________________

18. What percentage of the total energy use is devoted to:

a. transportation __________ percent
b. industry __________ percent
c. residential __________ percent
d. electrical power generation __________ percent

19. Match the dominant primary energy sources with the list of secondary energy uses.

a. oil-based fuels   b. natural gas   c. coal   d. nuclear power and other sources
e. more than one of these energy sources is used for the stated purpose

[ ] transportation
[ ] industrial processes
[ ] space heating and cooling
[ ] generation of electrical power

20. The energy source that supplies more than 40 percent of our total energy and virtually 100 percent of our energy for transportation is (a. oil-based fuels, b. natural gas, c. coal, d. nuclear power and other sources).

21. All nuclear power and virtually all coal is used to generate __________________ power.

The Exploitation of Crude Oil

22. We are now dependent on foreign sources for more than __________ percent of our crude oil

23. List four economic, political, or ecological costs of foreign oil dependence.

a. ______________________________ b. ______________________________

c. ______________________________ d. ______________________________

How Fossil Fuels Are Formed

24. Three examples of fossil fuels are:

a. ____________________ b. ____________________ c. ____________________

25. Fossil fuels are accumulations of organic matter from rapid ______________________ activity that occurred at various times in Earth's history.

26. Indicate whether the following statements are true [+] or false [-] concerning fossil fuels.

[ ] Supplies are limited.
[ ] The significant accumulations of organic matter that produced them can happen today.
[ ] We are using them far faster than they can be produced.

Crude Oil Reserves vs. Production

27. Amounts of crude oil that are estimated to exist in Earth are called ________________ reserves.

28. Oil fields that have been located and measured are called __________________ reserves.

29. The maximum annual production from an oil field is limited to about __________ percent of the (proven, remaining) reserves.

Assume that a field has 100 million barrels of recoverable reserves. Given a maximum production of 10 percent, how much can be withdrawn in

30. Year 1: ____________________ barrels leaving __________________ barrels

31. Year 2: ____________________ barrels leaving __________________ barrels

32. (True or False) Production from a given field may continue indefinitely.

33. (True or False) As maximum production goes down, the price per barrel will increase.

Declining U.S. Reserves and Increasing Importation

34. List two revelations of the Hubbart curve applied to U.S. oil production.

    a. ____

    b. ____

The Oil Crisis of the 1970s

35. Indicate whether the following events have increased [I] or decreased [D] in the United States since the 1970s.

    [ ] consumption of fuels derived from oil
    [ ] discoveries of new oil in the United States
    [ ] production of oil in the United States
    [ ] the gap between production and consumption
    [ ] United States dependence on foreign oil

36. OPEC stands for

    ____

37. The OPEC nations created an oil crisis in the 1970s by (increasing, suspending) oil exports.

38. Through its ability to create world oil shortages, OPEC (increased, decreased) the price of crude oil.

Adjusting to Higher Prices

39. Indicate whether increasing the price of crude oil caused the following events to increase [I] or decrease [D] in the United States.

    [ ] the rate of exploratory drilling and discovery of new oil
    [ ] renewed production from old oil fields
    [ ] efforts toward fuel conservation
    [ ] consumption
    [ ] development of alternative energy sources
    [ ] dependence on foreign oil

40. (True, False) The oil crisis of the 1970s created an oil glut.

41. As a result of oil surpluses, the price of a barrel of crude oil (increased, decreased).

42. As a result of oil surpluses, the price of a gallon of gas at the gas station.

    a. increased b. decreased c. stayed at the same level as during the crisis.

Victims of Our Success

43. Indicate whether the collapse in oil prices caused the following events to increase [I] or decrease [D] in the United States.

    [ ] the rate of exploratory drilling and discovery of new oil
    [ ] renewed production from old oil fields
    [ ] efforts toward fuel conservation
    [ ] consumption
    [ ] development of alternative energy sources
    [ ] dependence on foreign oil

44. U.S. dependence on foreign oil in 1994 has grown to __________ percent.

45. The single largest factor in the U.S. balance-of-trade deficit is __________________________..

Problems of Growing U.S. Dependency on Foreign Oil

46. List three problems resulting from U.S. dependency on foreign oil.

    a. ________________________________ b. ________________________________

    c. ________________________________

47. The cost for foreign crude oil represents about __________ percent of the U.S. balance trade deficit.

48. Compare the economic impact of using

    a. oil produced at home: ____

b. foreign oil: ____

49. The Persian Gulf War was fought primarily to

a. free the people of Kuwait
b. protect Kuwait oil fields from Iraq
c. drive oil prices up
d. force OPEC nations to come to terms on oil prices

50. Military intervention (increased, decreased) the price per barrel of oil.

51. Our actual cost of a barrel of oil from Persian Gulf sources is $35 + $__________ for military support services or $__________ a barrel.

52. The last major oil discovery in the United States was in ____________________ in 1968.

53. The Middle East possesses __________ percent of Earth's proven oil reserves (see Table 13-2).

54. List three ways we can reduce our dependency on foreign oil.

a. ____

b. ____

c. ____

Other Fossil Fuels

55. The United States is rich in supplies of fossil fuels, including:

a. ______________________ b. ____________________ c. ____________________

56. Identify the fossil fuel (a. natural gas, b. coal, c. synfuels, d. oil shales, e. oil sands) that ranks the highest in each of the following categories.

[ ] air pollution
[ ] cost of extraction
[ ] proven reserves
[ ] greenhouse effect
[ ] retrofitting to existing transportation technologies
[ ] habitat alteration
[ ] cost competition with current oil prices

Sustainable Energy Options

57. Which fossil fuel produces the least CO_2 emissions? ____________________

58. Identify two options to combat future oil shortages.

a. ______________________________ b. ______________________________

Conservation

59. A major preparation for future oil shortages is to (increase, decrease) oil consumption.

60. One way oil consumption can be decreased is through ________________.

61. (True or False) Between 1970 and 1980 the U.S. economy expanded without a corresponding rise in energy consumption.

62. At least ________ percent of the conservation reserve remains to be tapped.

63. List four ways to increase the conservation reserve.

a. ____

b. ____

c. ____

d. ____

Development of Non-Fossil Fuel Energy Sources

64. Identify two major pathways for developing alternative energy sources.

a. ____

b. ____

NO VOCABULARY DRILL FOR THIS CHAPTER!

SELF-TEST

1. Rank (1 to 5) the following fuel types in their order of historical use by human society.

[ ] coal [ ] wood [ ] electricity [ ] gasoline [ ] natural gas

Classify each energy type in terms of (a. primary or b. secondary) source and (a. renewable or b. nonrenewable).

Energy Types	Primary or Secondary	Renewable or Nonrenewable
2. wood	[ ]	[ ]
3. coal	[ ]	[ ]

Energy Types	Primary or Secondary	Renewable or Nonrenewable
4. oil-based	[ ]	[ ]
5. natural gas	[ ]	[ ]
6. electricity	[ ]	[ ]
7. nuclear	[ ]	[ ]
8. synfuels	[ ]	[ ]
9. solar	[ ]	[ ]

Classify each energy type in terms of primary end uses in the United States (a. transportation, b. industry, c. residential, d. electric power generation or e. not used).

10. [ ] wood
11. [ ] coal
12. [ ] oil-based fuels
13. [ ] natural gas
14. [ ] electricity
15. [ ] nuclear fuels
16. [ ] synfuels
17. [ ] solar energy

Indicate which energy types have all three advantages of low cost of extraction, low pollution, and high known reserves. (Check all that apply.)

18. [ ] wood
19. [ ] coal
20. [ ] oil-based fuels
21. [ ] natural gas
22. [ ] electricity
23. [ ] nuclear fuels
24. [ ] synfuels
25. [ ] solar energy

26. Which of the above energy types is a primary source, renewable, not used, and has all three advantages listed? ____________________

27. Explain why this energy source is not used.

____

____

Define a fossil fuel and give two examples.

28. Fossil fuel: ____

Two examples are: 29. ____________________ and 30. ____________________

Explain why fossil fuels are nonrenewable and nonrecyclable.

31. nonrenewable: ____

32. nonrecyclable: ____

33. Explain why electric heat systems represent a double energy waste:

____

34. Indicate whether the following actions will increase [+], decrease [-], or have no impact [0] on future supplies of oil-based fuels.

[ ] searching for new reserves
[ ] taking OPEC oil by force
[ ] increasing the price of gasoline at the pump
[ ] increasing use of mass transit systems
[ ] continuing consumption at present rates
[ ] car pooling
[ ] increasing engine fuel efficiency
[ ] increasing the beginning driving age to 18

35. Identify an oil field that has a potential production of 6 million barrels per day, is three times the size of the Alaskan field, is inexhaustible, and the exploitation of which will not adversely affect the environment.

This oil field is called the ______________________ reserve.

ANSWERS TO STUDY GUIDE QUESTIONS

1. human; 2. power; 3. steam; 4. b; 5. a; 6. 80; 7. internal combustion, oil, crude; 8. 39; 9. 22; 10. A, A, A, D; 11. secondary; 12. a; 13. A, D, D, D, D; 14. 30-40; 15. 60-70, heat, thermal; 16. major source of acid rain, flooding of terrestrial habits from dams, disposal of nuclear wastes; 17. transportation, industry, residential, electrical power generation; 18. 27, 39, 35, 36; 19. a, e, e, c; 20. a; 21. electrical; 22. 50; 23. trade imbalances, military actions, pollution of the oceans, coastal oil spills; 24. coal, crude oil, natural gas; 25. photosynthetic; 26. +, -, +; 27. estimated; 28. proven; 29. 10, remaining; 30. 10 million, 90 million; 31. 9 million, 81 million; 32. false; 33. true; 34. U.S. oil production peaked in 1970, will become increasingly dependent on foreign oil; 35. I, D, D, I, I; 36. Organization of Petroleum Exporting Countries; 37. suspending; 38. increased; 39. I, I, I, D, I, D; 40. true; 41. decreased; 42. c; 43. D, D, D, I, D, I; 44. 50; 45. purchase of foreign oil; 46. costs of purchases, supply disruption, resource limitations, 47. 26; 48. provides jobs that produce other goods and services, money leaves United States as another form of foreign aid; 49. b; 50. increase; 51. 56, 91; 52. Alaska; 53. 64; 54. use other fossil fuel sources, reduce our energy demand, develop alternatives to fossil fuels; 55. natural gas, coal, oil shales and tar sands; 56. b, c, b, b, a, a or b, e; 57. natural gas; 58. conservation, develop non-fossil fuel sources; 59. decrease; 60. conservation; 61. true; 62. 80; 63. increase fuel efficiency in cars, cogeneration, use florescent lights, increase home insulation; 64. nuclear and solar power

---

ANSWERS TO SELF-TEST

1. 2, 1, 4, 3, 5; 2. a, a; 3. a, b; 4. a, b; 5. a, b; 6. b, b; 7. a, b; 8. a, b; 9. a, a; 10. e; 11. d; 12. a; 13. b or c; 14. b or c; 15. d; 16. e; 17. e; 18 - 24. not checked; 25. checked; 26. solar; 27. a multitude of political, economic, and cultural reasons; 28. derived from ancient photosynthetic processes; 29. coal; 30. oil; 31. cannot be made again in the short term; 32. cannot be used again once burned; 33. burn coal to produce heat to turn a turbine to produce electricity to produce heat for a home; 34. 0, 0, +, +, -, +, +, +; 35. conservation

CHAPTER 14

NUCLEAR POWER: PROMISE AND PROBLEMS

Nuclear Power: Dream or Delusion?

1. In the 1960s and early 1970s, the perception of nuclear power plants was one of unguarded (optimism, pessimism).

2. Since 1975, the perception of nuclear power plants has been one of (optimism, pessimism).

3. What factor has the greatest impact on future developments of nuclear power plants?

   a. cost b. government financing c. public opinion d. foreign investments

4. What country produces 75 percent of its electricity from nuclear power (see Fig. 14-4)?

   a. United States b. Japan c. France d. Russian Republics

5. (True or False) Impending oil shortages are real.

How Nuclear Power Works

6. State the objective of nuclear power technology.

____

From Mass to Energy

7. Which of the following is a definition of [a] fission or [b] fusion?

   [ ] A large atom of one element is split into two smaller atoms of different elements.
   [ ] Two small atoms combine to form a larger atom of a different element.

8. In fission and fusion, the mass of the products is (more, less) than the mass of the starting material.

9. The lost mass is converted into tremendous amounts of ____________________.

10. Controlled fission releases this energy gradually as ____________________.

11. All nuclear power plants use (fission, fusion).

12. The raw material of nuclear power plants is uranium (235, 238).

13. Uranium-235 and 238 are both ________________ of the element called uranium.

14. Isotopes are defined by the number of (a. protons b. neutrons c. electrons) in an element.

15. The isotope that will fission is uranium (235, 238).

16. The "bullet" that starts the uranium-235 fission process is a (a. proton b. neutron c. electron d. element).

17. Three products of a uranium-235 fission are (see Fig. 14-5).

    a. __________________ b. __________________ c. __________________

18. A ____________________ reaction occurs when the neutrons cause other fissions, which release more neutrons, which cause other fissions.

19. The majority of all uranium is the (235, 238) isotope.

20. Enrichment is the process of enhancing the concentration of uranium (235, 238).

21. A nuclear bomb explosion is the result of a (controlled, uncontrolled) fission of a (high, low) grade of uranium 235.

22. A nuclear reactor is designed to:

a. sustain a continuous chain reaction (True or False)
b. prevent amplification into a nuclear explosion (True or False)
c. consist primarily of an array of fuel and control rods (True or False)
d. make some material intensely hot (True or False)
e. convert heat into steam (True or False)
f. convert steam into electricity (True or False)

23. Indicate whether the following are functions of [a] the moderator, [b] fuel rods or [c] control rods.

[ ] contain the mixture of uranium isotopes
[ ] contain a neutron-absorbing material
[ ] starts the chain reaction
[ ] controls the chain reaction
[ ] stores the heat energy to produce steam
[ ] slows down the fission neutrons

24. A nuclear reactor is simply an assembly of:

a. ________________ b. ________________ c. ________________

25. A nuclear power plant is designed to:

a. convert water into steam (True or False)
b. use steam to drive turbogenerators (True or False)
c. convert steam into electricity (True or False)
d. prevent meltdown (True or False)
e. produce super heated water in a reactor vessel (True or False)

26. A loss-of-coolant accident has the potential of causing ______________.

Comparisons of Nuclear Power with Coal Power

27. The fissioning of a pound of uranium fuel releases energy equivalent to burning ________ tons of coal.

28. Indicate whether the following characteristics pertain to [C] coal-fired or [N] nuclear power plants.

[ ] requires 3.5 million tons of raw fuel
[ ] requires 1.5 tons of raw material
[ ] will emit more than 10 million tons of carbon dioxide into the atmosphere
[ ] will emit no carbon dioxide into the atmosphere
[ ] will emit more than 400 thousand tons of sulfur dioxide into the atmosphere
[ ] will emit no acid-forming pollutants
[ ] will produce about 100 thousand tons of ash
[ ] will produce about 250 tons of radioactive wastes
[ ] presents the possibility of catastrophic accidents, e.g., meltdown

The Hazards and Cost of Nuclear Power

Radioactive Emissions

29. The fission products of unstable isotopes are called ________________.

30. Radioisotopes gain stability by ejecting ________________ particles or high-energy ________________.

31. Subatomic particles and high-energy radiation are referred to as ________________ emissions.

32. The indirect and direct products of fission are the ______________ wastes of nuclear power.

33. (True or False) Other materials in and around the reactor may become unstable isotopes.

34. Unstable isotopes are (direct, indirect) products of fission.

35. List five biological effects of radioactive emissions.

    a. ____

    b. ____

    c. ____

    d. ____

    e. ____

36. Radioactive emissions from nuclear power plants that are operating normally are generally (higher, lower) than background radiation.

37. List two sources of background radiation.

    a. ________________________ b. ________________________

38. Public exposure to radiation from normal operations of a power plant is < ________ percent of natural background.

39. List the two real public concerns regarding radiation emissions from nuclear plants.

    a. ________________________ b. ________________________

Radioactive Wastes

40. The process whereby unstable isotopes eject particles and radiation is known as ________________ decay.

41. The main variable that determines the rate of radioactive decay of a given isotope is its ____________________ life.

42. Half-lives vary from a fraction of a ____________________ to ____________________ of years.

43. Which of the following isotopes has the shortest [S] and which has the longest [L] half-life (see Table 14-2)?

    [ ] iodine-131, [ ] cesium-137, [ ] plutonium-239

44. Which of the above isotopes would require the

    a. shortest containment time? ________________

    b. longest containment time? ________________

45. Generally, (1, 10, 100) half-lives are required to reduce radiation levels to insignificant levels.

46. To be safe, plutonium-239 (half-life = 24,000 years) would require ___________ thousand years of containment.

47. Radioactive waste disposal falls into the two categories of ______________________ term and ______________________ term containment.

48. The major problems of radioactive waste disposal are related to

    a. finding long-term containment sites (True or False)
    b. transport of highly toxic radioactive waste across the United States (True or False)
    c. the lack of any resolution to the radioactive waste problem (True or False)

49. The world's commercial reactors have accumulated __________ thousand tons of nuclear wastes at the end of 2000.

50. List two types of nuclear wastes that present serious disposal problems.

    a. ______________________________ b. ______________________________

51. Of what significance are the following two locations concerning radioactive waste disposal?

    a. Savannah River: ____

    b. Lake Karachay: ____

52. (True or False) Dismantling a decommissioned power plant will generate more nuclear waste than the plant produced during its active life.

53. What is the significance of the acronym NIMBY in the radioactive disposal debate?

    ____

54. What area of the United States might become the nation's nuclear dump? _______________

55. State the purpose of the Nuclear Waste Policy Act of 1997.

____

56. Today, the majority of high-level nuclear waste disposal is in the form of (a. long-term storage facilities or b. on-site storage).

The Potential for Accidents

57. The catastrophic ramifications of the nuclear accident at Chernobyl were intensified due to the lack of a _____________________ building.

58. The near catastrophic nuclear accident at Three Mile Island was due to lack of _______________ in the reactor core.

59. In both the Chernobyl and Three Mile Island cases, what was the fundamental cause of both accidents?

____

60. Public confidence in nuclear engineering and engineers has (increased, decreased) since Chernobyl and Three Mile Island.

61. Nonsmokers are (equally, less, more) likely to suffer the effects of radiation as smokers. The answer is not in the book, but think about it!

Safety and Nuclear Power

62. Nuclear proponents now claim that nuclear reactors are __________ times safer.

63. Nuclear proponents are attempting to increase public confidence in nuclear power through:

a. active safety designs (True or False)
b. passive safety designs (True or False)
c. advanced light water reactors (True or False)
d. professing the very low probabilities of nuclear accidents (True or False)

Economic Problems with Nuclear Reactors

64. Utilities are turning away from nuclear power primarily because of:

a. adverse public opinion
b. government constraints
c. increased costs
d. concern for the environment

65. The cost of building a nuclear power plant has escalated due to:

a. new safety standards (True or False)

b. construction delays (True or False)
c. shorter-than-expected lifetimes of nuclear power plants (True or False)
d. withdrawal of government subsidies (True or False)
e. the demand for electricity being met by coal-fired power plants (True, False)

66. List two factors that have contributed to shorter-than-expected lifetimes of nuclear power plants.

a. ______________________________ b. ______________________________

More Advanced Reactors

67. Identify two alternative nuclear power technologies.

a. ______________________________ b. ______________________________

68. Indicate whether the following are characteristics of breeder [BR], fusion [FU], or both [BO] types of reactors.

[ ] creates more fuel than it consumes
[ ] the raw material is uranium-238
[ ] splits atoms
[ ] produces radioactive wastes
[ ] produces plutonium-239 as radioactive waste
[ ] fuses atoms
[ ] releases energy
[ ] the raw material is deuterium and tritium
[ ] the source of unprecedented thermal pollution

The Future of Nuclear Power

Opposition

69. Public opposition to nuclear energy is based on:

a. ____

b. ____

c. ____

d. ____

70. (True or False) Advances in nuclear power technology will help the energy needs of transportation.

Rebirth of Nuclear Power?

71. If worldwide nuclear plants were replaced by fossil-fuel power plants, CO_2 emissions would (rise or fall) by __________ percent.

72. Indicate whether the following conditions would [+] or would not [-] have to be met in order for nuclear power to be seriously considered as a future energy source.

[ ] greater reactor safety
[ ] reactor design
[ ] streamlining of licensing and monitoring
[ ] the political correctness of nuclear power
[ ] waste disposal
[ ] siting of new reactors

VOCABULARY DRILL

Directions: Match each term with the list of definitions and examples given in the tables below. Term definitions and examples might be used more than once.

Term	Definition	Example(s)	Term	Definition	Example(s)
active safety			half-life		
background radiation			indirect products		
breeder reactor			isotopes		
corrosion			LOCA		
direct products			LWRs		
embrittlement			mass number		
enrichment			passive safety		
fission			radioactive emissions		
fission products			radioactive decay		
fuel rods			radioactive wastes		
fusion			radioisotopes		
fusion reactor			rem		

VOCABULARY DEFINITIONS
a. a large atom of one element is split to produce two different smaller elements
b. two small atoms combine to form a larger atom of a different element
c. different forms of the same element
d. particles or rays emitted during fission
e. the number of neutrons and protons in the nucleus of the atom
f. process of producing a material containing a higher concentration of a desired element
g. rods full of uranium pellets
h. loss-of-coolant accident
i. unstable isotopes of the elements resulting from the fission process
j. materials that become radioactive by absorbing neutrons from the fission process
k. a measure of the ability for radioactive emission to do biological damage
l. radiation that occurs naturally in Earth's crust
m. occurs when unstable isotopes eject particles and radiation and become stable and nonradioactive
n. time for half the amount of a radioactive isotope to decay
o. safety that relies on operator-controlled actions
p. safety that relies on engineering devices and structure
q. effect on materials from constant neutron bombardment
r. effect on plumbing from corrosive chemicals
s. reactors that produce fuel in the process of consuming fuel
t. reactors that use H sources as fuel elements
u. reactors that use very pure water as the reactor moderator

VOCABULARY EXAMPLES	
a. cracked metal or pipes	i. Table 14-2
b. alpha and beta particles, neutrons	j. Superphenix
c. Fig. 14-8	k. Fig. 14-7
d. Fig. 14-5a	l. Fig. 14-14
e. Fig. 14-5b	m. humans
f. uranium-235	n. Table 14-1
g. Fig. 14-10	o. Three Mile Island
h. radon gas	

SELF-TEST

1. Which of the following factors will have the greatest impact on determining future developments of nuclear power?

   a. cost b. government financing c. public opinion d. foreign investments

2. Which of the following best describes nuclear fission?

   a. A large atom of one element is split into two smaller and different atoms with the release of free neutrons and energy.
   b. Two small atoms are melted together to form a larger atom with some mass being converted to energy.
   c. Energy is used to split a large atom into two smaller atoms with the release of energy.
   d. All of the above are proper explanations of fission.

3. The fuel for present nuclear power reactors is

   a. uranium-235 b. uranium-238 c. plutonium-239 d. thorium

4. The "bullet" that starts the uranium-235 fission process is

   a. a proton b. a neutron c. an electron d. an atom

5. What is the process called that results in the splitting of one uranium atom that releases free neutrons, causing other uranium atoms to split, which in turn releases more neutrons, causing other uranium atoms to split?

   a. nuclear fusion b. nuclear fission c. chain reaction d. enrichment

6. The process of enhancing the concentration of uranium-235 is called:

   a. nuclear fission b. a meltdown c. a chain reaction d. enrichment

7. Which of the following is not a function of the nuclear reactor?

   a. sustain a continuous chain reaction
   b. prevent amplification into a nuclear explosion
   c. convert heat into steam
   d. make some material intensely hot

8. Which of the following is not a function of the fuel rod?

   a. contain a mixture of uranium isotopes
   b. start a chain reaction
   c. control a chain reaction
   d. become intensely hot

9. A nuclear power plant is designed to

a. produce super heated water
b. convert heat into steam
c. convert steam into electricity
d. do all of the above

10. Which of the following is a disadvantage of nuclear power plants when compared to coal-fired power plants?

a. the amount of raw material required
b. the type of waste produced
c. the wattage of electrical power produced
d. the amount of sulfur dioxide produced

11. The major disadvantage(s) of using atomic energy to generate electricity is/are that

a. it causes a large amount of thermal pollution
b. it produces very dangerous waste products
c. the ultimate goal of putting a nuclear power plant on line is redundant to the goal of coal-fired generating plants
d. All of the above are disadvantages.

12. Radioactive emissions are not know to cause

a. damage to biological tissues
b. damage to DNA molecules
c. a greenhouse effect
d. increases in background radiation

13. The major problems of radioactive waste disposal are not related to

a. finding long-term containment facilities
b. transport of highly toxic radioactive wastes across the United States
c. a lack of resolution to the radioactive waste problem
d. background radiation around nuclear power plants

14. Escalating costs of building a nuclear power plant are not related to

a. environmental concerns
b. new safety standards
c. construction delays
d. embrittlement

15. The lack of public support for nuclear technology is based on

a. distrust of the technology and those who support it
b. resulting environmental pollution problems
c. their inability to understand the technology
d. all of the above

ANSWERS TO STUDY GUIDE QUESTIONS

1. optimism; 2. pessimism; 3. c; 4. c; 5. true; 6. control nuclear reactions so that energy is released gradually as heat; 7. a, b; 8. less; 9. energy; 10. heat; 11. fission; 12. 235; 13. isotopes; 14. b; 15. 235; 16. b; 17. radioactive by-products, heat, neutrons; 18. chain; 19. 238; 20. 235; 21. uncontrolled, high; 22. true, true, true, true, false, false; 23. b, c, b, c, a, a; 24. fuel elements, moderator-coolant, movable control rods; 25. all true; 26. meltdown; 27. 50; 28. C, N, C, N, C, N, C, N, N; 29. radioisotopes; 30. subatomic, radiation; 31. radioactive; 32. radioactive; 33. true; 34. indirect; 35. damage biological tissue, block cell division, death, damage DNA molecules, cancer, birth defects; 36. lower; 37. Earth's crust, cosmic rays from outer space; 38. 1; 39. disposal of radioactive waste, accidents; 40. radioactive; 41. half; 42. seconds, thousands; 43. [S] = iodine-131, [L] = plutonium-239; 44. iodine-131, plutonium-239; 45. 10; 46. 240; 47. short, long; 48. all true; 49. 180; 50. military and high-level nuclear wastes; 51. deliberate release of radioactive wastes, 20-year nuclear waste discharge area; 52. true; 53. the public response "not in my back yard"; 54. Yucca Mountain in southwestern Nevada; 55. requires the DOE to establish an interim nuclear waste storage facility (MRS); 56. b; 57. containment; 58. coolant; 59. human error; 60. decreased; 61. equally; 62. 6; 63. false, true, true, false; 64. c; 65. all true; 66. embrittlement, corrosion; 67. breeder and fusion reactors; 68. BR, BR, BR, BO, BR, FU, BO, FU, FU; 69. a general distrust of the technology, skepticism of management, overall safety of nuclear power plants, nuclear waste disposal problems; 70 false; 71. rise, 8; 72. all +

ANSWERS TO SELF-TEST

1. c; 2. a; 3. a; 4. b; 5. c; 6. d; 7. c; 8. c; 9. d; 10. b; 11. d; 12. c; 13. d; 14. a; 15. d

VOCABULARY DRILL ANSWERS

Term	Definition	Example(s)	Term	Definition	Example(s)
active safety	o	m	half-life	n	i
background radiation	l	h	indirect products	j	g
breeder reactor	s	j	isotopes	c	f
corrosion	r	a	LOCA	h	o
direct products	i	f	LWRs	u	c
embrittlement	q	a	mass number	e	f
enrichment	f	f	passive safety	p	l
fission	a	d	radioactive emissions	d	b
fission products	a	d	radioactive decay	m	i
fuel rods	g	k	radioactive wastes	j	g
fusion	b	e	radioisotopes	i	i
fusion reactor	t	e	rem	k	n

CHAPTER 15

RENEWABLE ENERGY

Principles of Solar Energy

1. Solar energy is a (potential, kinetic) form of energy originating from (chemical, thermonuclear) reactions in the sun.

2. Solar energy is a (renewable, nonrenewable) form of energy.

Putting Solar Energy to Work

3. The four main hurdles to overcome in using solar energy are:

   a. ______________________________ b. ______________________________

   c. ______________________________ d. ______________________________

4. List three direct uses of solar energy.

   a. ______________________________ b. ______________________________

   c. ______________________________

Solar Heating of Water and Solar Space Heating

5. Indicate whether the below components are required of active [A], passive [P], or both [B] types of solar heating systems (see Figs. 15-6 to 15-8).

   [ ] flat-plate solar collector
   [ ] water pump
   [ ] heat storage facility
   [ ] expansion tank
   [ ] blowers
   [ ] backup system
   [ ] heat exchanger
   [ ] valves
   [ ] improved insulation
   [ ] appropriate landscaping
   [ ] earth berms

6. Indicate whether the below components are requirements of solar hot water [W], space [S], or both [B] types of solar heating systems (see Figs. 15-6 to 15-8).

   [ ] flat-plate solar collector
   [ ] water pump
   [ ] heat storage facility
   [ ] expansion tank
   [ ] blowers
   [ ] backup system
   [ ] heat exchanger
   [ ] valves
   [ ] improved insulation
   [ ] appropriate landscaping
   [ ] earth berms

7. The most economical means of solar heating is an/a (active, passive) system.

8. A well-designed passive solar home can reduce energy bills by __________ percent with added construction costs of __________ to __________ percent.

9. (True or False) Partial conversion to solar energy will make a large amount of natural gas available for fueling vehicles.

10. About __________ percent of the total U.S. energy budget is used for heating buildings and water.

11. The chief barrier to more widespread use of passive solar designs is ________________.

Solar Production of Electricity

12. List two technologies that convert solar energy into electricity.

    a. ______________________ b. ______________________

13. A device that converts light directly to electricity is the __________________ cell.

14. The photovoltaic or solar cell consists of two layers of

    a. water b. electrons c. atoms d. mylar film

15. The top layer contains atoms with (additional, missing) electrons in the outer orbital.

16. The bottom layer contains atoms with (additional, missing) electrons in the outer orbital.

17. The top layer will readily (lose, accept) an electron.

18. The bottom layer will readily (lose, accept) an electron.

19. The kinetic energy of sunlight forces a movement of electrons from the top to the bottom layer creating an electrical imbalance or ___________________.

20. Solar cell technology:

    a. is becoming more cost-effective with increased use (True or False)
    b. will not wear out (True or False)
    c. is already competitive in costs per watt output (True or False)
    d. has the potential to provide power for anything from a watch to a home (True or False)

21. Solar trough collectors convert __________ percent of incoming sunlight to electric power at a cost of __________ cents per kilowatt-hour.

22. The economic feasibility of the solar trough collector is enhanced by (more, less) environmental pollution when compared to coal-fired facilities.

23. Power towers use mirrors to convert solar energy to _________________ which runs a _________________ that creates electricity.

24. The dish/engine system has demonstrated efficiencies of __________ percent.

The Promise of Solar Energy

25. List four disadvantages of using solar energy technologies.

    a. ____________________ b. ____________________

    c. ____________________ d. ____________________

26. Identify three hidden costs in the use of traditional energy sources.

    a. ____________________ b. ____________________

    c. ____________________

27. What is the best strategy for integrating solar energy technology with existing energy consumption patterns?

    ____

28. About __________ percent of our electrical power is generated by coal-burning and nuclear power plants.

29. Complete the following statements with MORE or LESS

    The use of solar electrical power will:

    a. create ________________ reliance on coal or nuclear power
    b. lead to significantly ________________ acid rain
    c. produce ________________ electrical power for villages in developing countries

Solar Production of Hydrogen -The Fuel of the Future

30. Indicate whether the following are benefits [+] or limitations [-] of hydrogen power.

    [ ] substitute for natural gas
    [ ] substitute for gasoline
    [ ] pollution factor
    [ ] production technology
    [ ] modification of existing transportation technologies
    [ ] production costs
    [ ] national distribution system

Indirect Solar Energy

31. List four indirect forms of solar energy.

    a. ____________________ b. ____________________

    c. ____________________ d. ____________________

Hydropower

32. The earliest form of hydropower was the use of (natural, artificial) water falls.

33. The contemporary form of hydropower is the use of (natural, artificial) water falls.

34. Artificial waterfalls are created by building huge ________________.

35. The artificial waterfalls of huge dams generate ________________ power.

36. Indicate whether the following are benefits [+] or drawbacks [-] of hydroelectric power.

[ ] level of pollution generated
[ ] level of environmental degradation
[ ] amount of total energy produced
[ ] geographical distribution of energy produced
[ ] ecological impacts above and below the dam

37. Only __________ percent of the nation's rivers remain free-flowing.

38. Many of the remaining free-flowing rivers are protected by the

____.

39. The Nam Theun Two Dam will displace ______________ people and destroy ________________ square miles of sensitive environments.

40. The Three Gorges Dam will displace ______________ people.

Wind Power

41. The earliest form of harnessing wind power were wind (mills, turbines).

42. The contemporary form of harnessing wind power is wind (mills, turbines).

43. Indicate whether the following are benefits [+] or drawbacks [-] of wind power.

[ ] the size limitations on wind turbines
[ ] the number of megawatts of electricity produced
[ ] the level of pollution generated
[ ] the level of environmental degradation
[ ] geographical distribution of energy produced
[ ] cost-effectiveness
[ ] aesthetics

Biomass Energy

44. List four examples of biomass energy or bioconversion.

a. ________________________ b. ________________________

c. ____________________________ d. ____________________________

45. Indicate whether the following are benefits [+] or drawbacks [-] of biomass energy.

[ ] availability of the biomass resource
[ ] access to the biomass resource
[ ] public acceptance and utilization of biomass energy
[ ] past history of human harvests within a maximum sustained yield

46. Methane gas is (direct, indirect) use of biomass as fuel.

47. (True or False) One can get more electrical power from cows than from nuclear power.

48. Alcohol production is a (direct, indirect) use of biomass as fuel.

49. Alcohol production utilizes (fermentation, distillation, both).

Additional Renewable Energy Options

50. List two additional sustainable energy options.

a. ____________________________ b. ____________________________

Geothermal Energy

51. Indicate whether the following are benefits [+] or drawbacks [-] of geothermal energy.

[ ] if accessible, it is an everlasting energy resource
[ ] the consistency in number of megawatts of electricity produced
[ ] the level of pollution generated
[ ] the level of environmental degradation
[ ] geographical distribution of energy produced
[ ] cost-effectiveness
[ ] technology required for extraction

Tidal Power

52. Tidal power converts the energy of (incoming, outgoing, both) tides.

53. Indicate whether the following are benefits [+] or drawbacks [-] of tidal power.

[ ] if accessible, it is an everlasting energy resource
[ ] the consistency in number of megawatts of electricity produced
[ ] the level of pollution generated
[ ] the level of environmental degradation
[ ] geographical distribution of energy produced
[ ] cost-effectiveness
[ ] technology required for extraction

Ocean Thermal Energy Conversion

54. Indicate whether the following are benefits [+] or drawbacks [-] of OTEC.

[ ] technology
[ ] cost-effectiveness
[ ] environmental consequences

Policy for a Sustainable Energy Future

55. Indicate whether the following are [+] or are not [-] aspects of the existing U.S. energy policy.

[ ] mandates to increase fuel efficiency of cars
[ ] exploitation of public lands for fossil fuel reserves
[ ] subsidies to producers of solar energy or solar energy products
[ ] continued research on and development of renewable energy resources
[ ] continued research on and development of alternative energy technologies
[ ] advocacy of energy conservation
[ ] public education on solar and other renewable energy sources

56. Indicate whether the following energy policies represent "business as usual" [B] or "transition to the future" [T].

[ ] tax subsidies for solar energy technology production and use
[ ] mandates to increase fuel efficiency of cars
[ ] depletion allowances
[ ] military support to assure access to oil in the Middle East
[ ] damage to human health
[ ] savings of $600 million in electric bills
[ ] cars that get 80 miles per gallon
[ ] climate change research
[ ] deregulation of the electrical industry
[ ] net metering
[ ] carbon tax

VOCABULARY DRILL

Directions: Match each term with the list of definitions and examples given in the tables below. Term definitions and examples might be used more than once.

Term	Definition	Example(s)	Term	Definition	Example(s)
active systems			OTEC		
bioconversion			passive systems		
biomass energy			power gird		
carbon tax			PV cell		
electrolysis			solar trough		
flat-plate collectors			wind farms		
fuel cells			wind turbine		
geothermal energy					

VOCABULARY DEFINITIONS
a. broad box with a glass top and a black bottom with imbedded water tubes
b. moves water or air with pumps or blowers
c. moves water or air with natural convection currents or gravity
d. cell that converts sunlight energy into electricity
e. a network of power lines taking power from generating stations to customers
f. long trough-shaped reflectors tilted toward the sun
g. disassociation of H_2O using electricity
h. devices in which H is recombined with O_2 chemically to produce electric potential
i. wind-driven generator
j. collections of thousands of wind-driven generators
k. deriving energy from present-day photosynthesis
l. concept of using ocean's temperature differences to produce power
m. naturally heated groundwater
n. tax on fuels according to the amount of CO_2 produced during consumption

VOCABULARY EXAMPLES	
a. burning wood	h. Hawaii
b. Fig. 15-12	i. Fig. 15-8
c. Fig. 15-18	j. pumps, valves, blowers
d. Fig. 15-6	k. European countries
e. Fig. 15-23	l. Fig. 15-15
f. overhead electrical lines	m. Fig. 15-20
g. H and O_2 bubbles	

SELF-TEST

1. The biggest problem connected with using solar energy is

a. there is not enough energy inherent in sunlight to be worth its use
b. there is no good means of collecting sunlight energy
c. the means of storing it are bulky and expensive
d. there are no good ways of converting solar energy into profits

Match the below terms or concepts with examples of solar and other renewable energy sources.

Terms and Concepts

2. [ ] flat-plate collector
3. [ ] active solar system
4. [ ] passive solar system
5. [ ] photovoltaic cells
6. [ ] solar trough collector
7. [ ] power tower
8. [ ] hydrogen production
9. [ ] hydropower
10. [ ] wind turbine
11. [ ] bioconversion
12. [ ] OTEC
13. [ ] geysers
14. [ ] twice-daily rise and fall

Examples

a. solar energy
b. additional renewable energy options
c. indirect solar energy

15. The most cost-effective method of using solar energy for home use is through

a. passive solar systems
b. active solar systems
c. photovoltaic solar systems
d. selling your excess power back to the utility companies

16. Which of the following is not a direct means of using solar energy?

a. direct conversion to electrical power
b. heating homes and other buildings
c. production of flammable hydrogen gas
d. production of biomass for conversion to flammable liquids

17. Solar cell technology

a. cannot become cost-effective even with increased use
b. will wear out
c. is competitive with nuclear power in costs per watt output
d. has severe limitations in power output

18. A major drawback of solar production of hydrogen is

a. its ability to replace our reliance on natural gas
b. its ability to replace gasoline in cars
c. its waste products
d. the ability to produce free hydrogen atoms

19. Hydroelectric power cannot provide a large percentage of power in the future because of the

a. level of thermal pollution generated
b. amount of energy produced
c. limited geographic distribution of energy produced
d. level of air and water pollution generated

20. A major drawback of wind power is the

a. amount of energy produced
b. size limitations of wind turbines
c. level of environmental degradation
d. cost-effectiveness

21. The major drawback that geothermal energy, tidal power and wave power all have in common is

a. the number of megawatts of electricity that could be produced is low
b. the technology required for energy conversion is not cost-effective
c. the level of environmental degradation
d. the limitations on long-term dependency as an energy source

22. In view of the nature of U.S. energy problems and associated environmental problems, which of the following areas should be given a higher priority in research and development funding?

a. fusion power b. solar power c. breeder reactors d. utilization of coal

ANSWERS TO STUDY GUIDE QUESTIONS

1. kinetic, thermonuclear; 2. renewable in that the sun will burn for a very long time; 3. collection, conversion, storage, cost-effectiveness; 4. heating, electricity, fuel; 5. B, A, A, A, A, B, A, A, B, B, P; 6. B, W, B, W, B, B, W, W, S, S, S; 7. passive; 8. 75, 5, 9. true; 10. 25; 11. ignorance; 12. photovoltaic cells, solar trough collectors; 13. solar; 14. b; 15. missing; 16. additional; 17. accept; 18. accept; 19. current; 20. true, false, false, true; 21. 22, 10; 22. less; 23. steam, turbogenerators; 24. 30; 25. expensive technologies, only works during the day, requires back-up energy source or storage battery, some climates not sunny enough; 26. air pollution, strip-mining, nuclear waste; 27. use it during the daytime hours to meet peak energy demands and use traditional energy sources in the evening hours; 28. 78; 29. less, less, more; 30. +, +, +, -, -, -,+; 31. wind, water, biomass, ocean thermal; 32. natural; 33. artificial; 34. dams; 35. hydroelectric; 36. +, -, -, -, -; 37. 2; 38. Wild and Scenic Rivers Act of 1968; 39. 4,500, 2,485; 40. 1.9 million; 41. mills; 42. turbines; 43. -, +, +, +, +, +, -; 44. fire wood, burning MSW, methane production, alcohol production; 45. +, -, +, -; 46. indirect; 47. true; 48. indirect; 49. both; 50. geothermal, tidal; 51. -, -, -, -, -, -, -; 52. both; 53. +, +, +, -, -, -; 54. -, -, +; 55. -, +, -, -, -, -, -; 56. T,T,B,B,B,T,T,T,T,T,T

ANSWERS TO SELF-TEST

1. d; 2. a; 3. a; 4. a; 5. a; 6. a; 7. a; 8. a; 9. c; 10. c; 11. c; 12. c; 13. b; 14. b; 15. a; 16. d; 17. d; 18. d; 19. c; 20. b; 21. b; 22. b

VOCABULARY DRILL ANSWERS

Term	Definition	Example(s)	Term	Definition	Example(s)
active systems	b	j	OTEC	l	h
bioconversion	k	a	passive systems	c	i
biomass energy	k	a	power gird	e	f
carbon tax	n	k	PV cell	d	b
electrolysis	g	g	solar trough	f	l
flat-plate collectors	a	d	wind farms	j	m
fuel cells	h	c	wind turbine	i	m
geothermal energy	m	e			

CHAPTER 16

ENVIRONMENTAL HAZARDS AND HUMAN HEALTH

1. The bacterium that causes Lyme disease comes from the ____________________, which can be transmitted to deer and humans by the primary vector called the ____________________.

2. Explain the relationship between oak trees, gypsy moths, mice, and Lyme disease.

   ____

   ____

3. A study of the connections between hazards in the environment and human disease and death is called ____________________ health.

Links Between Human Health and the Environment

4. In this chapter, environment means: ____

____.

5. In the context of environmental health, give three examples of hazards.

   a. ____________________ b. ____________________

   c. ____________________

6. Risk is defined as: ____

____.

The Picture of Health

7. List four dimensions of the meaning of health.

   a. ____________________ b. ____________________

   c. ____________________ d. ____________________

8. Human life expectancy has (increased, decreased) in the last 40 years.

9. List three advances that caused a change in human life expectancy in the last 40 years.

   a. ____________________ b. ____________________

   c. ____________________

10. Thirty-five percent of the mortality is developing countries is caused by ____________________ compared to __________ percent in developed countries.

Environmental Hazards

11. What are two ways to consider hazards to human health?

   a. ____________________ b. ____________________

12. Match the hazards on the left with their classifications on the right.

Hazards	Classifications
[ ] smoking	a. cultural
[ ] AIDS	b. biological
[ ] plague	c. physical
[ ] malaria	d. chemical
[ ] tornadoes	

Hazards	Classifications
[ ] floods	a. cultural
[ ] pesticides	b. biological
[ ] cleaning agents	c. physical
[ ] sunbathing	d. chemical
[ ] uses vectors	
[ ] building in a floodplain	
[ ] consume harmful drugs	
[ ] respiratory infections	
[ ] leaded paint	

13. Explain the relationship between AIDS and tuberculosis.

____

14. Give the common pathogens that cause:

a. diarrheal diseases: _______________________

b. malaria: _______________________

15. What insect is the vector of malaria? _______________

16. What part of the human anatomy is infected by malaria? _______________

17. List three symptoms of malaria.

a. _______________ b. _______________

c. _______________

18. What country has the greatest exposure to tornadoes? _______________

19. List four methods of exposure to chemical hazards.

a. _______________ b. _______________

c. _______________ d. _______________

20. Toxicity = _______________ of exposure + _______________ of a toxic substance.

21. List four potential effects of toxic chemicals on human health.

a. _______________ b. _______________

c. _______________ d. _______________

Cancer

22. Twenty-three percent of all U.S. deaths are traced to ________________________.

23. __________ percent of cancers can be traced to environmental causes.

Pathways of Risk

The Risks of Being Poor

24. Why is poverty the world's biggest killer?

    ____

    ____

25. (True or False) Wealth = health.

26. (True or False) People in developed countries tend to live longer.

27. List three reasons that explain the differences in time and cause of death between people in developed and developing countries.

    a. __________________________ b. __________________________

    c. __________________________

28. How does equitable income distribution affect the health of a nation?

    ____

The Cultural Risk of Smoking

29. Smoking is responsible for __________ percent of the cancer deaths in the United States.

30. Tobacco related illness is estimated to cost $__________ billion/year in lost work capacity and health care expenses.

31. Since 1964, the U.S. smoking population has dropped from 40 percent to __________ percent.

32. What is ETS? ____

33. (True or False) Nicotine is addictive.

34. (True or False) The United States continues to subsidize the tobacco industry.

Risk and Infectious Diseases

35. What does an epidemiologist do?

____

36. List two major causes of food and water contamination.

a. ____________________ b. ____________________

37. (True or False) Major outbreaks of diarrheal diseases occur only in developing countries.

38. In addition to malaria, list four more diseases known to be carried by mosquitoes.

a. ____________________ b. ____________________

c. ____________________ d. ____________________

39. What major problem has developed in the campaign to control mosquito populations?

____

40. Indicate whether the following activities have increased [+] or decreased [-] the incidence of malaria.

[ ] deforestation
[ ] irrigation
[ ] creation of dams
[ ] high intensity agriculture

Toxic Risk Pathways

41. (True or False) Air inside the home and workplace often contains much higher levels of pollutants than outdoor air.

42. Factors contributing to indoor air pollution include

a. products used indoors (True or False)
b. well insulated and sealed buildings (True or False)
c. duration of exposure to indoor pollutants (True or False)

43. List five types or sources of indoor air pollutants (see Fig. 16-13).

a. ____________________ b. ____________________

c. ____________________ d. ____________________

e. ____________________

Risk Assessment

44. Which of the following actions carry the greatest risk (see Table 16-3)?

a. being a policeman b. drinking alcohol c. smoking cigarettes

Risk Assessment by the EPA

45. Risk analysis includes

a. hazard assessment (True or False)
b. dose-response assessment (True or False)
c. exposure assessment (True or False)
d. risk characterization (True or False)

46. Epidemiological studies are used to make a cause and effect relationship between human health and (Past, Future) exposures to carcinogenic chemicals.

47. Animal tests are used to make a cause-and-effect relationship between human health and (past, future) exposures to carcinogenic chemicals.

48. List three objections that have been raised about animal testing.

a. ______________________________ b. ______________________________

c. ______________________________

49. Match the steps in risk analysis with the methods used to perform each step.

Step	Method
[ ] hazard assessment	a. final estimate of probability of fatal outcome from hazard
[ ] dose-response assessment	b. associates dose with incidence and severity of response
[ ] exposure assessment	c. uses epidemiological studies and animal tests
[ ] risk characterization	d. studies human groups with previous exposures

Risk Management

50. List the two steps in risk management.

a. ____

b. ____

51. Regulatory decisions are based on:

a. cost-benefit analysis b. risk-benefit analysis c. public preferences d. all of these

Risk Perception

52. How many of the public's top 11 environmental concerns are also on the EPA's list of top 11 risks (see Table 16-4)? __________

53. The public's **risk perception** is (the same as, different from) that of EPA scientists.

54. List the outrage factors that influence the public's risk perception.

 a. ____________________ b. ____________________ c. ____________________

 d. ____________________ e. ____________________ f. ____________________

 g. ____________________

55. It is (public concern or cost-benefit analysis) the drives public policy.

56. (True or False)Public concern for human health overrides concern for major ecological risks.

VOCABULARY DRILL

Directions: Match each term with the list of definitions and examples given in the tables below. Term definitions and examples might be used more than once.

Term	Definition	Example(s)	Term	Definition	Example(s)
distributive justice			morbidity		
dose			mortality		
dose-response			outrage		
ecological risks			risk		
environmental health			risk assessment		
epidemiology			risk characterization		
exposure			risk management		
hazard			risk perception		
hazard assessment			vector		

VOCABULARY DEFINITIONS
a. disease-carrying agent
b. connections between hazards in the environment and human disease and death
c. anything that causes sickness, shortens life, or contributes to human suffering
d. incidence of disease in a population
e. incidence of death in a population
f. study of the presence, distribution, and control of disease in populations
g. probability of suffering from injury, disease, or death
h. process of evaluating the risks associated with a particular hazard before taking the action
I. concentrations of the chemical in a test
j. incidence and severity of response when exposed to different doses of a chemical
k. identifying subjects, origin of chemical, dose and length of exposure
l. probability of a fatal outcome due to a hazard
m. examining evidence linking a potential hazard to its harmful effects
n. concerning nature's welfare
o. intuitive judgment about risks
p. additional public concerns other than fatalities
q. risk characterization of a hazard followed by regulatory decisions
r. making sure all stakeholders are involved in the decision-making process

VOCABULARY EXAMPLES	
a. mosquito	g. involuntary risks
b. Table 16-3	h. Table 16-2
c. tracking geographic distribution of HIV	i. toxic chemicals and cancer
d. everyone votes on the decision	j. Table 16-1
e. How many will die?	k. Table 16-4
f. Chernobyl, Russia	

SELF-TEST

Fill in the below table with the human disease(s) that the list of organisms are known to cause or carry.

Organisms	Disease(s)
1. black-legged tick	
2. white-tailed deer	
3. mosquito	
4. Salmonella	
5. Escherischia coli	
6. Borrelia burgdorferi	
7. Plasmodium	
8. rat	

Indicate whether the following are a. cultural, b. biological, c. physical, or d. chemical hazards.

9. [ ] earthquakes
10. [ ] cholera
11. [ ] sunburn
12. [ ] cleaning solutions

13. In the United States, about what percent of all deaths can be traced to cancer?

    a. 15 b. 25 c. 50 d. 75

14. Why is poverty the world's greatest killer?

    ____

15. (True or False) People in developed countries live longer than people in developing countries. Explain your answer.

    ____

16. The major single cause of cancer deaths in the United States is ________________.

17. The smoking population in the United States has (increased, decreased) over the past 40 years.

18. (True or False) There could be a resurgence of malaria in developing countries.

19. List four indoor air pollutants

a. ______________________ b. ______________________

c. ______________________ d. ______________________

The below definitions are examples of [a] hazard, [b] risk, [c] outrage.

20. [ ] People who have a choice will accept risk at a greater rate than those having no choice.
21. [ ] The probability of suffering injury or other loss as a result of exposure to hazard.
22. [ ] Anything that has the potential of causing suffering due to exposure.

Match the steps in risk analysis with the methods used to perform each step.

	Step	Method
23.	[ ] hazard assessment	a. final estimate of probability of fatal outcome from hazard
24.	[ ] dose-response assessment	b. associates dose with incidence and severity of response
25.	[ ] exposure assessment	c. uses epidemiological studies and animal tests
26.	[ ] risk characterization	d. studies human groups with previous exposures

ANSWERS TO STUDY GUIDE QUESTIONS

1. mouse, black-legged tick; 2. Mice control gypsy moths, but carry the vector of Lyme disease, which infects deer feeding on the acorns provided by mice control of gypsy moths, The choice is a healthy forest or healthy human population; 3. environmental; 4. the whole context of human life - physical, chemical, biological; 5. injury and disease, property damage, environmental decay; 6. probability of suffering from injury, disease, or death; 7. physical, mental, spiritual, emotional; 8. increased; 9. social, medical, economic; 10. infectious diseases, 6; 11. lack of access to resources, exposure to a hazard; 12. a, a, b, b, c, c, d, d, a, b, c, a, b, d; 13. HIV infection weakens the immune system, which allows TB to flourish; 14. Salmonella, Campytobacter, E. coli, Plasmodium; 15. mosquito; 16. red blood cells; 17. fever, chills, headaches, vomiting; 18. United States.; 19. eating, drinking, breathing, direct use; 20. time, dose; 21. impairment of immune system, infertility, brain impairment, birth defects; 22. cancer; 23. 25; 24. The poor have no access to health care resources and have a greater exposure to hazards; 25. true; 26. true; 27. education, nutrition, modernization; 28. overall improvement of quality of life leads to healthier society; 29. 30; 30. 200; 31. 26; 32. second-hand tobacco smoke; 33. true; 34. true; 35. traces geographic distribution of a disease, mode of transmission and consequences; 36. lack of sewage treatment and personal hygiene; 37. false; 38. yellow fever, dengue fever, elephantiasis, encephalitis; 39. mosquito and Plasmodium resistance to controls; 40. all +; 41. true; 42. all true; 43. see Fig. 16-13; 44. c; 45. all true; 46. past; 47. future; 48. rodents and humans have different responses to same dose, higher than normal doses given to rodents, people object to animal experiments; 49. c, b, d, a; 50. complete risk assessment, regulatory decision; 51. d; 52. 3; 53. different; 54. technological ignorance, lack of choice, memory of past hazards, overselling safety, morality, lack of control, fairness; 55. public concern; 56. true

ANSWERS TO SELF-TEST

1. Lyme; 2. Lyme; 3. malaria, yellow fever, encephalitis, elephantiasis; 4. diarrheal disease; 5. diarrheal disease; 6. Lyme; 7. malaria; 8. bubonic plague; 9. c; 10. b; 11. a; 12. d; 13. b; 14. lack of access to health care resources and greater exposure to hazards; 15. true, people in developed countries have greater access to health care resources and less exposure to hazards; 16. smoking; 17. decreased; 18. true; 19. paints, solvents, aerosols, pesticides, smoking; 20. a; 21. c; 22. b; 23. c; 24. b; 25. d; 26. a

---

VOCABULARY DRILL ANSWERS

Term	Definition	Example(s)	Term	Definition	Example(s)
distributive justice	r	d	morbidity	d	j
dose	i	f	mortality	e	j
dose-response	j	f	outrage	p	g
ecological risks	n	h	risk	g	b
environmental health	b	j	risk assessment	h	k
epidemiology	f	c	risk characterization	l	e
exposure	k	f	risk management	q	j
hazard	c	b	risk perception	o	j
hazard assessment	m	b	vector	a	a

CHAPTER 17

PESTS AND PEST CONTROL

The Need for Pest Control

Defining Pests

1. (True or False) The term pest refers only to bugs.

Match the organisms on the left with their pest category on the right.

	Organism	Pest Category
2.	[ ] weeds	a. disease
3.	[ ] molds	b. annoyance
4.	[ ] coyotes	c. competitive feeder
5.	[ ] snails	d. predator
6.	[ ] bacteria	e. decomposer
7.	[ ] mosquitoes	f. nutrient competitor

The Importance of Pest Control

8. (True or False) Humankind need no longer rely on pest control.

9. Crop losses to pests have (increased or decreased) since the 1950s.

10. An estimated _________ percent of agricultural production are lost to pests now compared to __________ percent in 1950s.

11. Indicate whether the following chemicals are meant to target [a] plants or [b] animals.

    [ ] pesticides [ ] herbicides

Different Philosophies of Pest Control

12. List the two philosophies of pest control.

    a. ______________________________ b. __________________________

13. Which philosophy treats only the symptoms of pest outbreaks? ____________________

14. Which philosophy stimulates ecosystem immunity to pest outbreaks? __________________

15. Which philosophy combines the two approaches? ____________________

Promises and Problems of the Chemical Approach

Development of Chemical Pesticides and Their Successes

16. Indicate whether the following chemicals are meant to target: [a] all organisms, [b] mice and rats, [c] insects, or [d] fungi.

    [ ] fungicides [ ] rodenticides [ ] insecticides [ ] biocides

17. A common chemical pesticide used in the 1930s was

    a. cyanide b. arsenic c. sulfur d. DDT

18. Indicate whether the following are characteristics of [a] first-generation, [b] second-generation, or [c] both types of pesticides.

[ ] consisted of toxic heavy metals
[ ] are inorganic compounds
[ ] are synthetic organic compounds
[ ] are persistent
[ ] promote pest resistance
[ ] are broad spectrum in their effect
[ ] examples are cyanide and arsenic
[ ] an example is DDT (dichlorodiphenyltrichloroethane)
[ ] stimulated Rachel Carson to write

Problems Stemming from Chemical Pesticide Use

19. List three problems associated with the use of synthetic organic pesticides.

a. ______________________________

b. ______________________________

c. ______________________________

20. Chemical pesticides gradually lose their ____________________.

21. Loss of effectiveness leads to two choices by growers, namely, use (more, less) or use the (same, new) pesticides.

22. The use of chemical pesticides on insects leads to selection for the (sensitivity, resistance) allele in the gene pool of the population.

23. Each generation of insects exposed to a chemical pesticide will have (more, less) alleles for resistance in the gene pool of the population.

24. (True or False) As a pest population becomes resistant to one pesticide, it may, at the same time, become resistant to others to which it has not yet been exposed.

25. Match the following definitions with [a] resurgence or [b] secondary pest outbreak.

[ ] insects that were previously not a problem become a pest category
[ ] the return of the pest at higher and more severe levels

26. Pesticide treatments often have a (less, greater) effect on the natural enemies of pests.

27. Which of the following statements are [+] or are not [-] explanations of why pesticides have a greater impact on natural enemies than upon the target pest?

[ ] Herbivore species are more resistant to the pesticides than are its predators.
[ ] Biomagnification gives the predator a higher dose of the pesticide.
[ ] Predatory species may be starved out due to a temporary lack of prey.

28. How is the chemical approach contrary to basic ecological principles?

____

____

29. How many million cases of acute occupational poisoning occur each year in developing countries? Between __________ and __________ million.

30. List three ways children and families in developing countries come into contact with pesticides.

a. ____________________ b. ____________________

c. ____________________

31. List three chronic effects of pesticide exposure on humans.

a. ____________________ b. ____________________

c. ____________________

32. What were three sources of evidence that indicated DDT was the causative factor in reproductive failure in fish-eating birds?

a. DDT levels in fragile egg shells were (high, low).
b. DDT (promotes, reduces) calcium metabolism.
c. DDT levels in fish-eating birds was (higher, lower).

33. The highest DDT levels can be expected at the (bottom, top) of the food chain.

34. DDT accumulates in the body ________________ of humans and other animals.

35. DDT was banned in the United States in the early 19__________s.

36. (True or False) When DDT was banned, pesticides were no longer used.

37. About __________ million tons of pesticides are used globally.

38. If less than 1 percent of the pesticides dumped on the environment come into contact with the pest organism, where does the remaining 99 percent of the pesticide go?

a. drifts into the air (True or False)
b. settles on surrounding ecosystems (True or False)
c. settles on bodies of water (True or False)
d. remains as residue on foods (True or False)
e. leaches from the soil into aquifers (True or False)
f. is captured and recycled by growers (True or False)

39. Beginning with the emergence of a pest problem, complete the steps in the pesticide treadmill below using [a] use of chemical pesticides, [b] resistance, resurgence, and secondary outbreaks, and [c] more pest problems (see Fig. 17-6).

Pest problems → [ ] → [ ] → [ ] → [ ]

Nonpersistent Pesticides: Are They the Answer?

40. Assume that 1000 pounds of DDT was applied to a field in Nebraska in 1950. How much of the 1000 pounds will be left in:

a. 1970 ______________ b. 1990 ______________ c. 2010 ______________

d. When will the DDT disappear? ______________

41. The half-life of nonpersistent pesticides is (a. days, b. weeks, c. years, d. a or b).

42. Indicate whether the following statements are true [+] or false [-] concerning nonpersistent pesticides.

[ ] stay in the environment for a long time
[ ] are less toxic than DDT
[ ] are broad spectrum pesticides
[ ] do not cause resistance in pest species
[ ] do not cause resurgence or secondary outbreaks

Alternative Pest Control Methods

43. List the four alternative pest control methods.

a. ______________________________

b. ______________________________

c. ______________________________

d. ______________________________

Cultural Controls

44. Indicate whether the following control measures are in the category of [a] sanitation or [b] personal hygiene.

[ ] water purification
[ ] bathing
[ ] wearing clean clothing
[ ] proper and systematic disposal of garbage
[ ] using clean cooking and eating utensils
[ ] refrigeration, freezing, and canning foods

45. Indicate whether the following control measures are examples of

a. selection of what to grow and where to grow it
b. management of lawns and pastures
c. water and fertilizer management
d. timing of planting
e. destruction of crop residues
f. bordering crops with beneficial non crop plants
g. crop rotation
h. polyculture
i. quarantines

[ ] prohibiting biological materials that carry pests from entering a country
[ ] planting plants under their optimum conditions
[ ] growing two or more species together or in alternate rows
[ ] cutting grass to a height no higher than 3 inches
[ ] alternating crops from one year to the next in a given field
[ ] providing the ideal levels of water and fertilizer
[ ] plowing under or burning the residue left after harvest
[ ] eliminating plants that act as pest attractants and grow others that act as repellents
[ ] delaying planting so most of the pest population starves before the plants are available

Control by Natural Enemies

46. Match the following examples of pests with the natural enemy that was used to control each pest.

Pest	Natural Enemy
[ ] scale insects	a. manatees
[ ] caterpillars	b. plant-eating insects
[ ] gypsy moths and Japanese beetles	c. bacteria
[ ] prickly pear cactus	d. parasitic wasps
[ ] water hyacinths	e. vedalia (ladybird) beetles
[ ] rabbits in Australia	

47. The first step in using natural enemies is _________________ of the natural enemies that already exist.

48. (True or False) Finding natural enemies to pests generates as much profit for industry as do synthetic chemicals.

49. Natural enemies are a (short, long) term form of pest control.

50. Natural enemies are (more, less) profitable in the long term than synthetic chemicals.

Genetic Control

51. List four forms of genetic controls.

a. ______________________________ b. ______________________________

c. ______________________ d. ______________________

52. A chemical barrier means that the plant produces some chemical that is ______________ or ______________ to potential pests.

53. A chemical barrier was introduced into wheat plants that killed Hessian fly ______________.

54. Physical barriers are ______________ traits that impede pest attacks.

55. An example of a physical barrier is ______________ hairs on leaf surfaces.

56. (True or False) Pests cannot overcome genetic controls through resistance.

57. The sterile male procedure involves subjecting the (a. egg, b. larvae, c. pupa, d. adults) to just enough high energy radiation to render them sterile.

58. (True or False) Pests cannot develop resistance to the sterile male technique.

59. An example of application of the sterile male technique is the ______________ fly.

60. What is a transgenic crop?____

____

61. Give two examples of transgenic crops.

a. ______________________ b. ______________________

62. List three drawbacks in the use of genetically altered crops.

a. ____

b. ____

c. ____

Natural Chemical Control

63. Indicate whether the following definitions pertain to [a] hormones or [b] pheromones.

[ ] a chemical secreted by one insect that influences the behavior of another
[ ] chemicals that provide signals for development and metabolic functions

64. The four aims of natural chemical control are to ______________, ______________, ______________, and ______________ an insect's own hormones or pheromones to disrupt its life cycle.

65. The two advantages of natural chemicals are that they are ______________ and ______________.

66. Juvenile hormone has a direct adverse affect on the (a. egg, b. larva, c. pupa, d. adult).

67. Sex attractant pheromones may be used in the ______________________ or ______________________ techniques.

Socioeconomic Issues in Pest Management

Pressures to Use Pesticides

68. Natural controls are aimed at (management, eradication) of pest populations.

69. Significant damage is implied when the cost of damage is considerably (greater, less) than the cost of pesticide application.

70. Indicate whether the following attitudes or actions are examples of:

a. cosmetic spraying   b. insurance spraying   c. chemical company vested interest

[ ] consumer desire for unblemished produce
[ ] the use of pesticides just to be safe
[ ] ignorance of whether a bug is causing damage or is beneficial
[ ] the Snow White syndrome
[ ] promoting and exploiting the above attitudes

Integrated Pest Management

71. Integrated pest management (IPM) addresses all ______________________, ______________________, and ______________________ factors.

72. IPM is a (short, long) term approach.

73. IPM (never, sometimes) requires use of chemical pesticides.

74. List five economic or environmental benefits derived from the Indonesian IPM program.

a. ____

b. ____

c. ____

d. ____

e. ____

Organically Grown Food

75. Control of pests in organic farming is likely to consist of (Check all that apply.)

[ ] nonpersistent pesticides
[ ] natural controls
[ ] cultural controls
[ ] synthetic organic pesticides

Public Policy

76. What are the three basic concerns in regulating the use of pesticides?

a. ______________________________

b. ______________________________

c. ______________________________

FIFRA

77. What is the purpose of FIFRA? ______________________________

78. The registration procedure requires ______________ to determine toxicity to animals.

79. List four shortcomings of FIFRA.

a. ______________________________

b. ______________________________

c. ______________________________

d. ______________________________

FQPA of 1996

80. What is the purpose of the Food Quality Protection Act of 1996?

____

____

Pesticides in Developing Countries

81. What is the annual export of pesticides to developing countries by the United States? ______________ metric tons.

82. (True or False) These pesticide exports do not include those banned in the United States.

83. Explain how PIC and FAO Codes of Conduct are protecting developing countries from imports of unwanted pesticides.

PIC: ____

FAO: ____

New Policy Needs

84. What new pesticide regulatory rules are needed to help farmers jump off the pesticide treadmill? (Check all that apply.)

[ ] groundwater protection
[ ] food safety
[ ] measuring extent and dosage of pesticide exposures
[ ] worker protection
[ ] expanded pesticide testing requirements

85. List one pesticide legislative reform that is being considered.

____

VOCABULARY DRILL

Directions: Match each term with the list of definitions and examples given in the tables below. Term definitions and examples might be used more than once.

Term	Definition	Example(s)	Term	Definition	Example(s)
biomagnification			natural chemicals		
broad-spectrum			nonpersistent		
chemical barriers			organically grown		
cosmetic spraying			persistent		
cultural control			pest		
economic threshold			pesticide		
FIFRA			pesticide treadmill		
first-generation			physical barriers		
genetic control			resistance		
herbicide			resurgence		
insect life cycle			second-generation		
insurance spraying			secondary outbreaks		
IPM			sex attractant		
natural enemies					

VOCABULARY DEFINITIONS
a. any organism that has a negative impact on human health or economics
b. chemicals that kill animal pests
c. chemicals that kill plants
d. pesticides used before the 1930s
e. pesticides used during and after the 1930s
f. pesticides that do not select target species only
g. pesticides that stay in the ecosystem for a long time
h. pest organism's ability to develop genetic immunity to a pesticide
i. pest population explodes to higher and more severe levels
j. population explosions of pests that previously were of no concern
k. the process of accumulating higher and higher doses through the food chain

VOCABULARY DEFINITIONS
l. cycle of pesticide use, resistance, resurgence, pesticide use, etc.
m. pesticides that remain for only a short period of time in the ecosystem
n. when economic losses due to pest damage outweigh cost of applying pesticide
o. use of pesticides to control pests that harm only item's outward appearance
p. use of pesticides when no control need is evident
q. metamorphic stages in an insect's life
r. chemical substance the alters the reproductive behavior of the opposite sex of the same species
s. identifying and using the natural predator of the pest species
t. developing a genetic incompatibility between the pest and its host species
u. alteration of one or more environmental factors that attract pest species
v. alteration of one or more stages in a pest species life cycle
w. substances that are lethal or repulsive to would-be pests
x. structural traits that impede the attack of a pest
y. involves the integration of several sequential pest control techniques
y. key legislation to control pesticides in the United States
z. crop production without the use of synthetic chemical pesticides or fertilizers

VOCABULARY EXAMPLES	
a. flies, tick, mosquitoes	m. just to be safe
b. DDT	n. Fig. 17-9
c. Round-up	o. pheromones
d. arsenic, cyanide	p. Fig. 17-11
e. Fig 17-4	q. sterile males
f. all pest species	r. Fig. 17-10
g. cotton mites	s. skunk odor
h. 0.001→ 0.01 → 0.10 → 1.0 ppm	t. hairy leaves
i. Fig. 17-6	u. Federal Insecticide, Fungicide, and Rodenticide Act
j. parathion	v. Integrated Pest Management
k. Fig. 17-15	w. Fig. 17-18

SELF-TEST

1. Define the term **pest**. ______________________________

2. Identify whether the following characteristics pertain to [a] first-generation, [b] second-generation, [c] both first- and second-generation, [d] nonpersistent, or [e] all pesticides.

   [ ] are broad spectrum in their effect
   [ ] are harmful to humans and the environment
   [ ] cause pest resistance
   [ ] cause resurgence
   [ ] cause secondary outbreaks

If a pesticide has been present in an aquatic ecosystem for several years, match the pesticide levels given below with various parts of an aquatic ecosystem food chain.

Pesticide Levels: a. 0.02 ppm, b. 2000 ppm, c. 40-300 ppm, d. 5 ppm, e. 1600 ppm

3. [ ] water 4. [ ] fish-eating birds 5. [ ] algae 6. [ ] plant-eating fish

7. Give the term that is used to explain your answers to 3 through 6. ______________________

What are three sources of evidence that implicated DDT as the causative factor in reproductive failure in fish-eating birds?

8. ______________________________

9. ______________________________

10. ______________________________

11. Explain the association between resistance and resurgence.

____

____

List four ways of controlling pest species that do not include the use of inorganic or synthetic organic chemicals.

12. ______________________ 13. ______________________

14. ______________________ 15. ______________________

Match the following pest control methods with your answers to questions 12 through 15.

16. using high energy radiation ______________________

17. crop rotation ______________________

18. use of pheromone ____________________

19. using ladybug beetles ____________________

20. chemical barriers ____________________

21. disposal of garbage ____________________

22. juvenile hormone ____________________

23. destruction of crop residues ____________________

24. Ecological pest management seeks to control pests by

a. manipulating natural factors affecting the host and pest relationship
b. letting nature take its course
c. using pesticides that are known to be ecologically sound
d. using biological controls only

25. Which of the following shortcomings in FIFRA promotes heavy lobbying by the chemical industry?

a. inadequate testing
b. lack of public input
c. pesticide exports
d. ban issued on a case-by-case basis when threats are proven

26. Cosmetic spraying is done to

a. improve the appearance of produce
b. improve the taste
c. improve the storability
d. protect the public health

ANSWERS TO STUDY GUIDE QUESTIONS

1. false; 2. f; 3. e, 4. d; 5. c; 6. a; 7. b; 8. false; 9. increased; 10. 37, 31; 11. b, a; 12. chemical technology, ecological pest management; 13. chemical technology; 14. ecological pest management; 15. integrated pest management; 16. d, b, c, a; 17. a; 18. a, a, b, c, c, c, a, b, b; 19. pest resistance, pest resurgence and secondary outbreaks, adverse environmental and human health effects; 20. effectiveness; 21. more, new; 22. resistance; 23. more; 24. true; 25. b, a; 26. greater; 27. -, +, +; 28. assumes ecosystem is a static entity in which one species, the pest, can be eliminated; 29. 3.5 and 5; 30. aerial spraying, dumping of pesticide wastes, using pesticide containers for drinking water storage; 31. dermatitis, cancers, neurological disorders, sterility, birth defects; 32. high, reduces, higher; 33. top; 34. fat; 35. 70; 36. false; 37. 3; 38. true, true, true, true, true, false; 39. a, b, a, b; 40. 500, 250, 125, never; 41. d; 42. -, -, +, -, -; 43. cultural, natural enemies, genetic, natural chemicals; 44. a, b, b, a, a, a; 45. i, a, h, b, g, c, e, f, d; 46. e, d, c, b, a, c; 47. conservation; 48. true; 49. long; 50. more; 51. chemical barriers, physical barriers, sterile males, strategies using biotechnology; 52. lethal, repulsive; 53. larva; 54. structural; 55. hooked; 56. false; 57. d; 58. true; 59. screw worm; 60. crops that have had the genes of other plant species, bacteria, or viruses incorporated into their genome; 61. Monsanto's Bollgard cotton, Roundup ready soybeans; 62. not well suited to developing countries, possible development of superweeds, development of insect resistance to transgenic crops; 63. b, a; 64. isolate, identify, synthesize, use; 65. specific, nontoxic; 66. b; 67. trapping, confusion; 68. management; 69. greater; 70. a, b, b, a, c; 71. social, economic, ecological; 72. long; 73. sometimes; 74. savings on pesticide purchases, no investments in pesticide applicators, environmental health, fish in rice paddies, health benefits; 75. blank, check the rest; 76. intended use of pesticides and impacts on human and environmental health, training of pesticide applicators, protection of food supply; 77. law requiring manufacturers to register pesticides with the government before marketing them; 79. inadequate testing, bans on case-by-case basis, pesticide exports, lack of public input; 80. establishes standards to protect food supply from becoming contaminated with high pesticide levels; 81. 200,000; 82. false; 83. exporting countries inform all potential importing countries of actions they have taken to ban or restrict use of pesticides, conditions of safe use of pesticides are addressed in detail; 84. all checked; 85. 50 percent reduction in pesticide use

---

ANSWERS TO SELF-TEST

1. any organism that is noxious, destructive, or troublesome; 2. all e; 3. a; 4. b; 5. d; 6. c; 7. biomagnification; 8. DDT levels in fragile egg shells was high; 9. DDT is known to reduce calcium metabolism; 10. DDT levels in fish-eating birds was higher than other organisms in the food chain; 11. pests develop a genetic resistance to pesticides exemplified by surviving members of the population surviving and reproducing many more pests (resurgence) that are resistant 12. natural enemies; 13. genetic controls; 14. natural chemicals; 15. cultural controls; 16. genetic control; 17. cultural control; 18. natural chemicals;194. natural enemies; 20. genetic controls; 21. cultural control; 22. natural chemicals; 23. cultural control; 24. a; 25. b; 26. a

VOCABULARY DRILL ANSWERS					
Term	**Definition**	**Example(s)**	**Term**	**Definition**	**Example(s)**
biomagnification	k	h	natural chemicals	v	o
broad-spectrum	f	b	nonpersistent	m	j
chemical barriers	w	h	organically grown	z	w
cosmetic spraying	o	l	persistent	g	b
cultural control	u	r	pest	a	a
economic threshold	n	k	pesticide	b	b
FIFRA	y	u	pesticide treadmill	l	i
first-generation	d	d	physical barriers	x	t
genetic control	t	q	resistance	h	e
herbicide	c	c	resurgence	i	f
insect life cycle	q	n	second-generation	e	b
insurance spraying	p	m	secondary outbreaks	j	g
IPM	y	v	sex attractant	r	o
natural enemies	s	p			

CHAPTER 18

WATER: POLLUTION AND PREVENTION

1. Loss of the sea grasses in the Chesapeake Bay was due to (a. toxic chemicals, b. herbicides, c. turbid water).

2. Using the terms MORE and LESS, compare the following characteristics of the bay ecology BEFORE and AFTER the massive die-off of sea grasses.

Ecological Characteristics	Before	After
a. depth of light penetration	______	______
b. amount of sediment	______	______
c. amount of phytoplankton	______	______
d. dissolved oxygen	______	______
e. bacteria	______	______

Ecological Characteristics	Before	After
f. photosynthesis	______	______
g. fish populations	______	______

Water Pollution

3. The presence of a substance in the environment that causes undesirable environmental alterations is called _________________.

Pollution Essentials

4. Match the sources of pollutants on the right with categories of pollutants on the left (see Fig. 18-1).

Categories	Sources
[ ] toxic chemicals	a. soil erosion and dredging
[ ] nutrient oversupply	b. refuse burning
[ ] sediments	c. coal-burning power plants
[ ] particulates	d. cars, trucks, buses, airplanes
[ ] acid forming compounds	e. leaching from lawns and agricultural fields
[ ] fuel combustion products	f. industrial discharges and disposal sites
[ ] pesticides/herbicides	g. fertilizer runoff from several sources

5. List four undesirable environmental alterations caused by pollution.

a. ___________________________ b. ___________________________

c. ___________________________ d. ___________________________

6. List the four actions required to reduce pollution problems.

a. ____

b. ____

c. ____

d. ____

7. What is one strategy that could be used to address pollution? ____________________

Water Pollution: Sources, Types, Criteria

8. Identify the following are point [P] or nonpoint [NP] pollution sources.

[ ] atmospheric deposition
[ ] oil wells
[ ] sewage systems
[ ] agricultural runoff

9. List two basic strategies that are employed to bring water pollution under control.

a. ______________________ b. ______________________

10. List four basic types of water pollutants.

a. ______________________ b. ______________________

c. ______________________ d. ______________________

11. The major problem resulting from untreated sewage wastes is the spread of infectious ______________.

12. Disease causing bacteria, viruses, and other parasitic organisms that infect humans and other animals are called ______________.

13. List four diseases caused by the pathogens present in sewage (see Table 18-1).

a. ______________________ b. ______________________

c. ______________________ d. ______________________

14. Transmission of pathogens in infected sewage waste may occur in what three ways?

a. ______________________ b. ______________________

c. ______________________

15. (True or False) Serious and widespread cholera outbreaks are still occurring.

16. Public health measures which prevent a disease cycle involve:

a. ____

b. ____

c. ____

17. Bacterial decomposition of organic matter in water will (increase or decrease) O_2 concentration.

18. A common water quality test that measures the amount of organic material, in terms of the amount of O_2 required to break it down biologically, is the ______________.

19. Classify the below list of pollutants as inorganic [IN] or organic [O].

[ ] heavy metals
[ ] petroleum products
[ ] pesticides
[ ] acid precipitation

20. The source of all sediments is ________________ erosion.

21. List six major points of soil erosion and sediments.

a. ____________________ b. ____________________ c. ____________________

d. ____________________ e. ____________________ f. ____________________

22. It is (True or False) that streams, rivers, lakes, and estuaries support complex ecosystems based on many kinds of plant and animal organisms living on or attached to the bottom.

23. Indicate whether the following types of damage to aquatic ecosystems is caused by [a] clay or [b] the bedload of sand and silt.

[ ] reduction of light penetration and rate of photosynthesis
[ ] smothers fish by clogging their gills
[ ] scours bottom-dwelling organisms clinging to rocks
[ ] fills in hiding and resting places for fish and crayfish
[ ] causes the stream to become more shallow

24. ______________________ is considered the foremost pollution problem of streams and rivers.

25. The most obvious point sources of excessive nutrients are __________________ outfalls.

26. The most notorious nonpoint source of excessive nutrients is __________________ runoff

27. The most obvious environmental impact of excessive nutrients is ________________.

28. It is estimated that _________ percent of our rivers, lakes, and estuaries are not meeting water-quality standards.

Eutrophication

Different Kinds of Aquatic Plants

29. Two kinds of aquatic plants are __________________________ plants, which are rooted to the bottom and __________________________ which mostly consist of single cells or small groups of cells that grow at or near the water surface.

30. List two kinds of benthic plants.

a. __________________________ b. ______________________

31. List two kinds of phytoplankton.

a. __________________________ b. ______________________

32. Phytoplankton can maintain itself near the surface where there is plenty of light, but they depend on (surface, bottom) nutrients.

33. Benthic plants get their nutrients from the (surface, bottom).

34. The balance between benthic plants and phytoplankton depends on the level of ________________________ in the water.

35. (Benthic plants or Phytoplankton) depend on light penetration to the bottom to enable photosynthesis.

The Impacts of Nutrient Enrichment

36. Indicate whether the following properties increase [+] or decrease [-] given the listed conditions.

    [+] nutrient → [ ] phytoplankton → [ ] turbidity → [ ] SAVs

37. The term oligotrophic means nutrient (rich, p oor).

38. The two filters of nitrogen and phosphate for natural waterways are:

    a. ________________________________ b. ________________________________

39. The major source of oxygen in aquatic ecosystems comes from the

    a. atmosphere b. benthic plants c. phytoplankton

40. Indicate whether the following conditions are LOW [-] or HIGH [+] in oligotrophic bodies of water. (You will need to read further into the chapter to answer all items.)

    a. [ ] dissolved oxygen b. [ ] bacterial decomposition

    c. [ ] light penetration d. [ ] benthic plants

    e. [ ] phytoplankton f. [ ] water temperature

    g. [ ] water turbidity h. [ ] nutrient concentrations

    i. [ ] species diversity j. [ ] aesthetic qualities

    k. [ ] recreational qualities l. [ ] BOD

    m. [ ] detritus decomposition n. [ ] sediments

41. The term eutrophic means nutrient (rich, poor).

42. Phytoplankton do not contribute significantly to dissolved oxygen because

    a. they do not carry on photosynthesis (True or False)
    b. they do not produce oxygen in photosynthesis (True or False)
    c. the oxygen they produce escapes from the surface into the atmosphere (True or False)

43. The amount of detritus (increases or decreases) in eutrophic water.

44. The increase in detritus comes from the dieback of (benthic plants, phytoplankton).

45. Attack of detritus by ________________________ uses so much ________________________ that bottom-welling fish and shellfish suffocate.

46. Decomposers (bacteria) can keep the dissolved oxygen at or near zero for as long as there is detritus to feed on,because in the absence of oxygen they can carry on ________________ respiration.

47. Indicate whether the following conditions are LOW [-] or HIGH [+] in eutrophic bodies of water. (You will need to read further into the chapter to answer all items.)

a. [ ] dissolved oxygen
b. [ ] bacterial decomposition
c. [ ] light penetration
d. [ ] benthic plants
e. [ ] phytoplankton
f. [ ] water temperature
g. [ ] water turbidity
h. [ ] nutrient concentrations
i. [ ] species diversity
j. [ ] aesthetic qualities
k. [ ] recreational qualities
l. [ ] BOD
m. [ ] detritus decomposition
n. [ ] sediments

48. A eutrophic body of water is not technically "dead" because it still supports the abundant growth of ________________ and ________________.

49. In shallow lakes and ponds, the lack of (SAVs, phytoplankton) are the indicators of eutrophication.

50. (True or False) Eutrophication of natural bodies of water is part of the natural aging process.

51. List three nutrient sources the cause cultural eutrophication.

a. ____________________ b. ____________________ c. ____________________

Combating Eutrophication

52. List the two general approaches to combating eutrophication.

a. ____

b. ____

53. List four methods that have been used in attacking the symptoms of eutrophication and state the major reason that each is less that successful or practical.

Method	Shortcoming
a. ______________________	______________________
b. ______________________	______________________
c. ______________________	______________________
d. ______________________	______________________

54. (True or False) The real control of eutrophication requires decreasing nutrient and sediment inputs.

55. List three major sources of nutrients that end up in streams and rivers.

a. ______________________

b. ______________________

c. ______________________

56. List seven long-term strategies for controlling eutrophication.

a. ______________________

b. ______________________

c. ______________________

d. ______________________

e. ______________________

f. ______________________

g. ______________________

57. The key to controlling eutrophication is to reduce ______________ inputs.

Sewage Management and Treatment

Development of Collection and Treatment Systems

58. Urban communities have two waste water collection and dispersal mechanisms. Which one

a. collects and drains runoff from precipitation? ______________

b. receives waste water from sinks, toilets and tubs? ______________

59. What person showed the relation between sewage-borne bacteria and infectious diseases?

____

60. The earliest forms of sewage disposal transferred it directly into ____________________ drains.

61. ____________________ societies initiated the system of flushing sewage into natural waterways.

62. Around 1900, sewage treatment included ____________________ the wastewater and separating ____________________ and ____________________ water.

63. (True or False) There are no longer any cases of raw sewage overflowing with storm water into waterways.

The Pollutants in Raw Sewage

64. The total of all the drainage from sinks, bathtubs, laundry, and toilets constitutes a mixture called raw ____________________.

65. Raw sewage is actually ____________ percent water and ____________ percent pollutants.

66. List the four major categories of pollutants in raw sewage and give an example of each.

Category	Example
a. ____________________	____________________
b. ____________________	____________________
c. ____________________	____________________
d. ____________________	____________________

Removing the Pollutants from Sewage

67. The two major sewage treatment problems in the United States are

a. ____________________ and b. ____________________

68. Match the following list of wastewater treatment technologies or products with [a] preliminary treatment, [b] primary treatment, [c] secondary treatment, including biological nutrient removal (BNR).

[ ] primary clarifiers
[ ] bar screen
[ ] grit settling tank
[ ] raw sludge
[ ] biological treatment
[ ] activated sludge system
[ ] activated sludge

[ ] rotating screens
[ ] course sand and gravel
[ ] removal of dissolved nutrients
[ ] uses decomposers and detritus feeders
[ ] settling of raw sludge
[ ] aeration

69. First, large pieces of debris are removed by passing the water through a _________________ screen.

70. Coarse sand and other dirt particles are removed by passing the water through a ______________________ settling tank where the velocity of the water is (increased, decreased) and the particles are allowed to ______________________ out.

71. The large debris from the bar screen are taken out and _________________.

72. Grit from the grit settling tank is put in _____________________.

73. In primary treatment

a. the water flows into very large tanks called ________________________ clarifiers.
b. the water flows through (quickly, slowly).
c. heavier particles of organic matter (disperse, settle out).

74. __________ percent of the organic material is removed by the settling process.

75. The material that settles out is called raw __________________________.

76. Secondary treatment is also called _____________________ treatment because decomposers and detritus feeders are used.

77. In the activated sludge system, a mixture of organisms is added to the water to be treated and it is passed through a tank ,where it is ________________________.

78. In the aeration tank, the organisms feed on __________________ matter.

79. Following the aeration tank, the water is passed on to a secondary ____________________ tank where the organisms are returned to the ____________________ tank.

80. Through the activated sludge systems, ____________ percent of the organic material is removed.

81. BNR utilizes aspects of the ___________________ and ____________________ cycles.

82. Indicate whether the following statements represent positive [+] or negative [-] attributes of using chlorine gas as the means of killing pathogenic bacteria in wastewater.

[ ] cost
[ ] toxicity during transportation and to fish.
[ ] chlorinated hydrocarbons.

83. Two alternative technologies for killing pathogenic bacteria in wastewater are

a. ______________________ and b. ______________________

Sludge Treatment

84. Raw sludge is about __________ percent water and __________ percent organic matter.

85. Three sludge treatment methods are

a. ______________________________ b. ______________________________

c. ______________________________

86. Anaerobic digestion of raw sludge in the absence of ___________________ produces the products of ________________ gas and ________________ sludge.

87. Composting basically produces ____________________.

88. Pasteurization and drying produce sludge __________________.

Alternative Treatment Systems

89. List three alternative methods of extracting wastewater nutrients.

a. ______________________________ b. ______________________________

c. ______________________________

Public Policy

90. List six indicators of progress in water pollution abatement since the enactment of the Clean Water Act of 1972.

a. ______________________________ b. ______________________________

c. ______________________________ d. ______________________________

e. ______________________________ f. ______________________________

VOCABULARY DRILL

Directions: Match each term with the list of definitions and examples given in the tables below. Term definitions and examples might be used more than once.

Term	Definition	Example(s)	Term	Definition	Example(s)
activated sludge			point-source		
bar screen			pollutant		
benthic plants			pollution		
biogas			preliminary treatment		
biological treatment			primary clarifiers		
BNR			primary treatment		
BOD			raw sewage		
chlorinated hydrocarbons			raw sludge		
co-composting			red tides		
composting toilet			SAV		
cultural eutrophication			sanitary sewers		
emergent vegetation			secondary treatment		
euphotic zone			sediment trap		
eutrophication			settling tank		
nonpoint source			sludge cake		
nonbiodegradable			sludge digester		
nontidal wetlands			storm drains		
oligotrophic			tidal wetlands		
pasteurized			treated sludge		
pathogens			turbid		
phytoplankton			wetlands		

VOCABULARY DEFINITIONS
a. human-caused addition of any material or energy in amounts that cause undesired alterations
b. material that causes pollution
c. resist attack and breakdown by detritus feeders and decomposers
d. nutrient rich
e. cloudy or murky
f. free-floating aquatic plants
g. deep rooted aquatic plants
h. aquatic plants that grow entirely under water
i. aquatic plants that grow partially under and above water
j. depth of adequate light for photosynthesis
k. nutrient poor
l. measure of pollution level by ppm oxygen required for decomposition
m. accelerated eutrophication caused by humans
n. land areas naturally covered by water for specified periods of time
o. coastal wetlands
p. inland wetlands
q. indicators of oceanic eutrophication
r. a fixed-point pollutant source
s. a diffuse pollutant source
t. a means of controlling erosion from construction sites
u. sewage with no treatment whatsoever
v. disease-causing organisms
w. system to collect runoff from precipitation
x. system to collect all wastewater
y. first stage in the removal of pollutants from sewage
z. device to remove large objects from wastewater
aa. device to remove sand and gravel from wastewater
bb. first stage in the removal of organic matter from wastewater
cc. device that allows organic matter to settle out of wastewater
dd. material removed during primary treatment
ee. stage in wastewater treatment that uses natural decomposers and detritus feeders
ff. a mixture of sludge and detritus-feeding organisms

VOCABULARY DEFINITIONS
gg. stage in wastewater treatment that removes dissolved nutrients
hh. result of spontaneous reactions between chlorine and organic compounds
ii. device that allows bacteria to feed on detritus in the absence of oxygen
jj. a major by-product from the sludge digester
kk. a dehydrated form of treated sludge
ll. a mixture of raw sludge and wood chips
mm. procedure for killing pathogens
nn. a waterless toilet

VOCABULARY EXAMPLES	
a. cattails, water sedges	p. methane
b. ppm dissolved O_2	q. Fig.18-14
c. lawn fertilizers and pet droppings	r. cancer-causing compounds
d. 1 inch to 600 feet	s. Fig. 8-15
e. Chesapeake Bay	t. Cilvus Multrum
f. urban runoff	u. Fig.18-13
g. synthetic organic compounds	v. Milorganite
h. swamps	w. bacteria and viruses
i. wetlands	x. fluid that enters the sewage treatment plant
j. filamentous or single-cell algae	y. plumbing from sinks, tubs, and toilets
k. sewage treatment plant	z. bacteria, protozoa, fungi
l. nitrate and phosphate	aa. Fig. 8-11
m. eutrophication	bb. street curb and gutter
n. marine phytoplankton	cc. nutrient-rich humus like material
o. hay bales on highway construction projects	dd. Florida Everglades

SELF-TEST

Indicate whether the following conditions are characteristic of [O] oligotrophic or [E] eutrophic bodies of water.

1. [ ] high dissolved oxygen
2. [ ] low light penetration
3. [ ] abundant phytoplankton
4. [ ] low water turbidity
5. [ ] high species diversity
6. [ ] high recreational qualities
7. [ ] high detritus decomposition
8. [ ] high bacterial decomposition
9. [ ] abundant benthic plants
10. [ ] cool water temperature
11. [ ] high nutrient concentrations
12. [ ] poor aesthetic qualities
13. [ ] low BOD
14. [ ] low sediments

List four major sources of sediments and nutrients that end up in streams and rivers.

Sediments	Nutrients
15. ____________________	19. ____________________
16. ____________________	20. ____________________
17. ____________________	21. ____________________
181. ____________________	22. ____________________

Explain how each of the following events or materials contribute to eutrophication.

23. Sedimentation: ____

____

24. Nutrients: ____

____

25. Decomposition: ____

____

26. Which of the following approaches to combating eutrophication is getting at the root cause?

a. chemical treatments b. decreasing nutrient input c. aeration d. harvesting algae

27. The foremost pollution problem of streams and rivers is

a. toxic wastes b. salinization c. sedimentation d. eutrophication

28. Nutrients flow into aquatic ecosystems from

a. fertilizers from croplands
b. animal wastes from feedlots
c. detergents containing phosphate
d. all of the above

29. Lands that are naturally covered by shallow water at certain times and more or less drained at others are called

a. grasslands b. wetlands c. rain forests d. estuaries

Indicate whether the following functions of wetlands will increase [+] or decrease [-] after the establishment of human "development" projects on or adjacent to wetlands.

30. [ ] water purification

31. [ ] provision of habitat and food for waterfowl and other wildlife

32. [ ] groundwater recharge

33. The two major problems resulting from untreated sewage wastes going into waterways are

a. ________________________ and b. ________________________.

34. The primary categories of pollutants in raw sewage are

a. pathogens, detritus, and dissolved nutrients
b. phosphates, nitrates, and sulfates
c. heavy metals, synthetic organisms, and pesticides
d. industrial, household, and municipal wastes

35. What proportion of raw sewage is actually pollutant? _________ percent

36. Match the following list of wastewater treatment technologies or products with [a] preliminary treatment, [b] primary treatment, or [c] secondary treatment, including BNR.

[ ] sand, grit, gravel
[ ] bar screen
[ ] aeration
[ ] raw sludge
[ ] denitrifying bacteria
[ ] activated sludge

37. Match each of the sewage treatment components [a to d] with the functions it performs.

a. activated sludge system b. bar screen c. primary clarifier d. none of these

[ ] screens out large pieces of debris
[ ] enables microorganisms to digest organic matter present
[ ] allows 50 to 70 percent of the organic matter to settle out
[ ] removes most of the dissolved nutrients

38. Most sewage treatment plants do not have

a. pretreatment b. primary treatment c. secondary treatment d. BNR

39. Explain why conventional methods of wastewater treatment only partially solve the problem of water pollution.

____

40. The treated water from most sewage treatment plants contains

a. nutrients b. bacteria c. sediments d. colloidal materials

ANSWERS TO STUDY GUIDE QUESTIONS

1. c; 2. more, less; less, more; less, more; more, less; less, more; more, less; more, less; 3. pollution; 4. f, g, a, b, c, d, e; 5. aesthetic, biological, human health, local to global; 6. identifying material or materials causing the pollution problem, identifying the source(s), develop and implement control strategies, avoid the pollution altogether; 7. apply the second principle of ecosystem sustainability; 8. NP, P, P, NP; 9. reduce the sources, treat the water to remove the pollutants; 10. pathogens, organic wastes, chemical pollutants, sediments, nutrients; 11. diseases; 12. pathogens; 13. see Table 18-1; 14. drinking, eating food, body contact; 15. true; 16. disinfection of public water supplies, improving personal hygiene and sanitation, sanitary collection and treatment of sewage wastes; 17. decrease; 18. biological oxygen demand (BOD test); 19. IN, O, O, IN; 20. soil; 21. croplands, overgrazed rangelands, deforested areas, construction sites, surface mining, gully erosion; 22. true; 23. a, a, b, b, b; 24. sediment; 25. sewage; 26. agricultural; 27. eutrophication; 28. 40; 29. benthic, phytoplankton; 30. submerged aquatic vegetation (SAVs), emergent vegetation; 31. single-cell bluegreen or filamentous green algae or diatoms; 32. surface; 33. bottom; 34. nutrients; 35. benthic plants; 36. +, +, -; 37. poor; 38. watersheds, wetlands; 39. b; 40. +, -, +, +, -, -, -, -, +, +, +, -, -, -; 41. rich; 42. false, false, true; 43. increases; 44. benthic plants; 45. decomposers, oxygen; 46. anaerobic; 47. -, +, -, -, +, +, +, +, -, -, -, +, +, +; 48. phytoplankton, fish; 49. SAVs; 50. true; 51. sewage treatment plants, poor farming practices, urban runoff; 52. attack the symptoms, get at the root cause; 53. (herbicide treatments, phytoplankton resistance), (aeration, expense), (harvesting algae, expense), (drawing water down, kills rooted aquatic plants); 54. true; 55. agriculture/forestry, urban/suburban runoff, sewage effluents; 56. banning phosphate detergents, upgrade sewage treatment plants, controlling agricultural and urban runoff, controlling sediments from construction and mining sites, controlling stream erosion, protecting wetlands, controlling air pollution; 57. nutrient; 58. storm drains, sanitary sewer line; 59. Pasteur; 60. storm; 61. Western; 62. treating, waste, storm; 63. false; 64. sewage; 65. 99.9, 0.1; 66. (debris and grit, rags and plastic bags), (particulate organic material, food wastes), (colloidal or organic material, feces), (dissolved inorganic materials, nitrogen); 67. nutrients, sludge; 68. b, a, a, b, c, c, c, c, a, c, c, b, c; 69. bar; 70. grit, decreased, settles; 71. incinerated; 72. land fills; 73. primary, slowly, settles out; 74. 30 to 50; 75. sludge; 76. biological; 77. aerated; 78. organic; 79. clarifier, aeration; 80. 90 to 95; 81. nitrogen, phosphate; 82. +, -, -; 83. ozone, ultraviolet radiation; 84. 98, 2; 85. anaerobic, composting, pasteurization; 86. oxygen, methane, treated; 87. humus; 88. pellets; 89. septic systems, using effluents for irrigation, reconstructing wetlands; 90. number of people served by adequate sewage treatment plants; reduction in soil erosion; more waterways safe for fishing and swimming; rivers, lakes and bays cleaned up and restored; fish returned to rivers; increase in bottom vegetation in Chesapeake Bay

---

ANSWERS TO SELF-TEST

1. O; 2. E; 3. E, 4. O; 5. O; 6. O; 7. E; 8. E; 9. O; 10. O; 11. E; 12. E; 13. O; 14. O; 15. croplands; 16. overgrazed rangelands; 17. deforested areas; 18. construction sites; 19. fertilizer from cropland; 20. fertilizer from lawns and gardens; 21. animal wastes from feedlots; 22. pet wastes; and human excrement, phosphate detergents, acid precipitation; 23. decreases light penetration to benthic plants; 24. increases phytoplankton growth; 25. decreases oxygen concentration killing animals; 26. b; 27. c; 28. d; 29. b; 30. -; 31. -; 32. -; 33. disease, eutrophication; 34. a; 35. 0.1 percent; 36. a, a, c, b, c, c; 37. b, a, c, a; 38. d; 39. most do not remove dissolved nutrients; 40. a

VOCABULARY DRILL ANSWERS					
Term	**Definition**	**Example(s)**	**Term**	**Definition**	**Example(s)**
activated sludge	ff	aa	point-source	r	k
bar screen	z	aa	pollutant	b	l
benthic plants	g	a	pollution	a	m
biogas	jj	p	preliminary treatment	aa	aa
biological treatment	gg	u	primary clarifiers	cc	aa
BNR	gg	u	primary treatment	bb	aa
BOD	l	b	raw sewage	u	x
chlorinated hydrocarbons	hh	r	raw sludge	dd	aa
co-composting	ll	no example	red tides	q	n
composting toilet	nn	t	SAV	h	a
cultural eutrophication	m	f	sanitary sewers	x	y
emergent vegetation	i	a	secondary treatment	gg	aa
euphotic zone	j	d	sediment trap	t	o
eutrophication	d	e	settling tank	cc	aa
non-point source	s	f	sludge cake	kk	s
nonbiodegradable	c	g	sludge digester	ii	g
nontidal wetlands	p	h	storm drains	w	bb
oligotrophic	k	i	tidal wetlands	o	e, dd
pasteurized	mm	v	treated sludge	jj	s, v
pathogens	v	w	turbid	e	e
phytoplankton	f	j	wetlands	n	e, h, dd

CHAPTER 19

MUNICIPAL SOLID WASTE: DISPOSAL AND RECOVERY

The Solid Waste Problem

Disposing of Municipal Solid Waste

1. List four factors that have contributed to ever-increasing amounts of MSW in the United States.

 a. ______________________ b. ______________________

 c. ______________________ d. ______________________

2. The three largest components of municipal solid waste are ______________________, ______________________, and ______________________ (see Fig. 19-2).

3. The proportions of the components of refuse change in terms of

 a. the source b. the neighborhood c. time of year d. all of these factors

4. Who assumes responsibility for collecting and disposing of municipal solid wastes?

____

5. The early form of municipal solid waste disposal was open ____________________, which were later converted to ____________________.

6. What proportions of the MSW in the United States are (see Fig. 19-3):

   a. dumped in landfills? _________ b. recycled? _________ c. combusted? _________

Improving Landfills

7. Indicate whether the following examples are related to the landfill problems of [a] leachate generation, [b] methane production, [c] incomplete decomposition, or [d] settling.

   [ ] explosions
   [ ] groundwater contamination
   [ ] development of shallow depressions
   [ ] a witch's brew
   [ ] biogas
   [ ] 30-year-old newspapers

8. Indicate whether the following landfill regulations address [a] leachate generation, [b] methane production, or [c] both a and b.

   [ ] siting landfills on high ground, well above the water table
   [ ] contoured floors to drain water into tiles
   [ ] covering the floor of the landfill with 12 inches of impervious clay or a plastic liner
   [ ] a gravel layer that surrounds the entire fill
   [ ] shaping the refuse pile into a pyramid
   [ ] monitoring wells

9. One of the most formidable problems facing city governments is ________________.

10. What is the implication of *NIMBY*?

____

11. List two undesirable consequences of siting.

   a. __________________________ b. __________________________

12. List two positive aspects of the siting problem.

   a. __________________________ b. __________________________

Combustion: Waste to Energy

13. Indicate whether the following are advantages [A] or disadvantages [D] of combustion.

[ ] 90 percent reduction in waste volume
[ ] generation of electricity
[ ] incinerator effluents
[ ] cost of construction
[ ] competition with recycling efforts
[ ] WTE
[ ] concrete blocks
[ ] resource recovery

14. Indicate whether the following are advantages [A] or disadvantages [D] of WTE facilities.

[ ] tons of MSW processed
[ ] fuel oil savings
[ ] recycling opportunities
[ ] resource recovery

Costs of Municipal Solid Waste Disposal

15. Increasing costs of disposing of MSW are related to

a. new design features of landfills (True or False)
b. acquiring landfill sites (True or False)
c. transportation to landfills (True or False)
d. tipping fees at landfills (True or False)

16. Acquiring a landfill site far enough away from suburbia automatically increases ____________________ costs.

17. Indicate whether the following events may [+] or may not [-] happen because of difficulties in establishing new landfill locations.

[ ] continued use of old landfills with inadequate safeguards
[ ] more old landfills closed than can be replaced with new landfills
[ ] a landscape covered with pyramids
[ ] reduced refuse production by society

Solutions

18. List three strategies to reduce waste at its source.

a. ____________________ b. ____________________ c. ____________________

Source Reduction

19. Match the methods of source reduction on the left with an example on the right.

METHODS	EXAMPLES
weight reduction	a. eliminating junk mail
the Information Age	b. paying more for products that last longer
reuse of durable goods	c. organizing garage sales
lengthening product life	d. making steel cans 60 percent lighter
volume reduction	e. shopping on the Internet

The Recycling Solution

20. What percent of the MSW is recyclable? __________ percent

21. List the two levels of recycling.

a. ________________________ b. ________________________

22. Match the following types of refuse with the type of product that reprocessing will produce.

Refuse Type	Product
[ ] paper	a. compost
[ ] glass	b. refabrication without ore extraction
[ ] plastic	c. synthetic lumber
[ ] metals	d. substitute for gravel and sand
[ ] food and yard wastes	e. cellulose insulation
[ ] textiles	f. strengthening agent in recycled paper

23. List the six characteristics of an effective municipal recycling program.

a. ____

b. ____

c. ____

d. ____

e. ____

f. ____

24. What proportion of the MSW stream is newspaper? __________ percent

25. (True or False) More paper is being recycled than landfilled.

26. Nonreturnable containers constitute what percentage of

a. solid waste stream in the United States? __________
b. nonburnable portion of MSW? ___________
c. roadside litter? ___________

27. Larger profits are generated from nonreturnable containers through

a. reduced transportation costs (True or False)
b. indefinite bottle production (True or False)
c. eliminating the returnable bottle competition (True or False)
d. opposing bottle laws (True or False)

28. Bottle laws encourage the use of (returnable, nonreturnable) containers.

29. Bottle laws will (decrease, increase) jobs and (decrease, increase) litter.

30. How many states have bottle laws? ___________

31. Who are the greatest opponents to bottle bills? ______________________________

32. (True or False) All plastic containers are recyclable.

33. List the products made from recycled plastic

PETE: ____

HDPE: ____

Composting

34. Composting produces a residue of decomposition called _________________.

35. (True or False) There are business opportunities in composting.

Public Policy and Waste Management

The Regulatory Perspective

36. Match the federal legislation with the role each plays in addressing MSW management.

a. Solid Waste Disposal Act of 1965
b. Resource Recovery Act of 1970
c. Resource Conservation and Recovery Act of 1976
d. Superfund Act of 1980
e. Hazardous and Solid Waste Amendments of 1984

[ ] gave EPA responsibility to set solid-waste criteria for all hazardous-waste facilities
[ ] gave jurisdiction over solid waste to Bureau of Solid Waste Management
[ ] encouraged states to develop some kind of waste management program
[ ] gave EPA power to close local dumps and set regulations for landfills
[ ] addressed abandoned hazardous waste sites

Integrated Waste Management

37. Integrated waste management uses (one, several) methods of handling MSW.

38. List three recommendations for MSW management that offer the greatest opportunity to promote sustainability.

a. ______________________ b. ______________________

c. ______________________

VOCABULARY DRILL

Directions: Match each term with the list of definitions and examples given in the tables below. Term definitions and examples might be used more than once.

Term	Definition	Example(s)	Term	Definition	Example(s)
landfill			resource recovery		
MSW			secondary recycling		
primary recycling			waste-to-energy		

VOCABULARY DEFINITIONS
a. surface repositories for MSW
b. conversion of MSW to electricity
c. separating and recovering materials from MSW
d. modern resource recovery facility
e. original waste material made back into the same material
f. total of all the materials thrown away from homes and commercial establishments
g. waste materials made into different products

VOCABULARY EXAMPLES	
a. Fig. 19-4	d. trash, refuse, garbage
b. Fig. 19-7	e. newspapers to cardboard
c. newspapers to newsprint	

SELF-TEST

1. Which of the following has not been a cause of increasing volumes of MSW?

   a. changing lifestyles b. disposable materials c. recycling d. increased gross national product

2. The three largest categories of municipal solid wastes are

   a. paper, cans, and bottles.
   b. paper, yard waste, metals
   c. plastics, paper, and metals
   d. food, paper, and plastics

3. The proportions of the components of refuse change in terms of

   a. the generator b. the neighborhood c. time of year d. all of these

4. Most municipal solid waste in the United States is presently disposed of

   a. by burning it in a closed incinerator
   b. in landfills
   c. by barging it to sea and dumping it
   d. in open-burning dumps

5. The potential of groundwater contamination from landfills is primarily due to

   a. leachate b. methane gas c. settling d. explosions

6. Newspaper into newsprint is an example of (primary, secondary) recycling.

7. Increasing costs of landfilling are related to

   a. new design features b. acquiring new sites c. transportation d. all of these

8. Even though establishing new landfill locations is extremely difficult, which of the following events will not occur in spite of this difficulty?

   a. continued use of old landfills with inadequate safeguards
   b. more old landfills closed than can be replaced with new landfills
   c. reduced refuse production by society
   d. a landscape covered with pyramids

9. Converting PETE into carpet fiber is an example of (primary, secondary) recycling?

10. In terms of recycling potential, paper can be

a. repulped and made into paper and paper products
b. manufactured into cellulose insulation
c. composted to make a nutrient-rich humus
d. all of the above

11. Which of the following examples of recycling laws places the burden mainly on the consumer rather than the government?

a. mandatory recycling
b. advance disposal fees
c. bans on the disposal of certain items
d. mandating government purchase of recycled materials

12. Burning raw refuse does not eliminate which of the following problems of recycling?

a. sorting b. hidden costs c. reprocessing d. marketing

13. Which of the following is not a potential value derived from incinerated refuse?

a. metal salvage b. returnable bottles c. extended use of existing landfills
d. fill dirt for construction sites

14. Which of the following is not a potential value of bottle bills?

a. reduced transportation costs
b. indefinite bottle production
c. increased jobs
d. decreased litter

15. Which of the following measures reduce the amount of material going into trash?

a. buying less b. decreased obsolescence c. yard sales d. all of these

ANSWERS TO STUDY GUIDE QUESTIONS

1. increasing population, changing lifestyles, disposable materials, excessive packaging; 2. paper, yard wastes, metals; 3. d; 4. government; 5. dumps, landfills; 6. 55, 28, 17; 7. b, a, d, a, b, c; 8. all a; 9. siting; 10. no one wants a landfill near their home or neighborhood; 11. increased cost of waste disposal, long-distance transfer of MSW; 12. residents reduce waste output, recycling occurs; 13. A, A, D, D, D, A, A, A; 14. A, A, D, A; 15. all true; 16. transportation; 17. +, +, +, -; 18. source reduction, recycling, composting; 19. d, e, c, b, a; 20. 75; 21. primary, secondary; 22. e, d, c, b, a, f; 23. strong incentive to recycle, recycling not optional, residential recycling is curbside, recycling goals are clear, involves local industries, municipality employs knowledgeable coordinator; 24. 38; 25. true; 26. 6, 50, 90; 27. all true; 28. returnable; 29. increase, decrease; 30. 10; 31. industry; 32. false; 33. carpet fibers, irrigation drainage tiles; 34. humus; 35. true; 36. e, a, b, c, d; 37. several; 38. waste reduction, waste disposal, recycling and reuse

---

ANSWERS TO SELF-TEST

1. c; 2. b; 3. d; 4. b; 5. a; 6. primary; 7. d; 8. c; 9. secondary; 10. d; 11. b; 12. b; 13. b; 14. b; 15. d

VOCABULARY DRILL ANSWERS

Term	Definition	Example(s)	Term	Definition	Example(s)
landfill	a	a	resource recovery	c	b
MSW	f	d	secondary recycling	g	e
primary recycling	e	c	waste-to-energy	b, d	b

CHAPTER 20

HAZARDOUS CHEMICALS: POLLUTION AND PREVENTION

Toxicology and Chemical Hazards

1. The study of harmful effects of chemicals on human and environmental health is called ________________________.

Dose Response and Threshold

2. Which portion of the figure represents

   [ ] the zone of NO harmful effect?
   [ ] the threshold level?
   [ ] the zone of harmful effects?

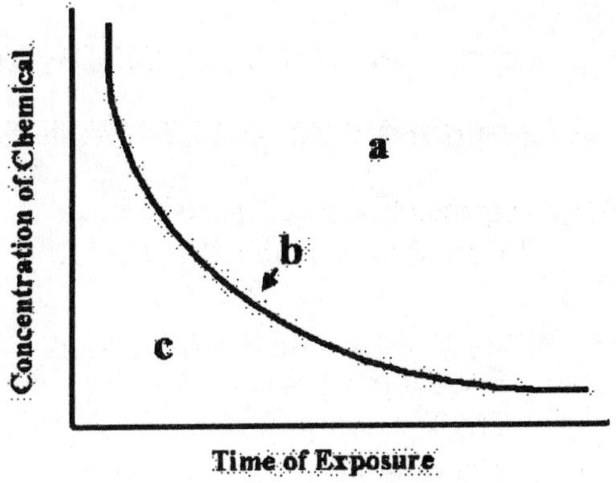

3. As the exposure time increases, the threshold level for harmful effects of toxic pollutants is (higher, lower).

4. The threshold level for different chemicals is the (same, different).

5. The threshold level for different people is the (same, different).

The Nature of Chemical Hazards: HAZMATs

6. Match the examples of hazardous materials on the left with the EPA category on the right.

[ ] gasoline	a. toxicity
[ ] acids	b. reactivity
[ ] explosives	c. corrosivity
[ ] pesticides	d. ignitability

Sources of Chemicals Entering the Environment

7. List five environmental costs in the product life cycle of your shampoo.

   a. ________________________________

   b. ________________________________

   c. ________________________________

   d. ________________________________

   e. ________________________________

8. (True or False) The predominant source of chemicals entering the environment are from households.

9. What is the purpose of the Toxics Release Inventory (TRI)?

   ____

The Threat from Toxic Chemicals

10. The two major classes of chemicals that are particularly significant environmental pollutants are ______________________________ and ______________________________.

11. Examples of heavy metals are:

    a. ____________________ b. ____________________ c. ____________________

    d. ____________________ e. ____________________ f. ____________________

    g. ____________________ h. ____________________

12. How are heavy metals used in industry?

    ____

13. Heavy metals are extremely toxic because they combine with and inhibit the functioning of ____________________, causing ____________________ and ____________________ effects.

14. Synthetic organic chemicals are the basis for the production of all ________________, synthetic ____________________, synthetic ____________________, ____________________ coatings, ____________________, ____________________, and ____________________ preservatives.

15. Synthetic organic chemicals are (biodegradable, nonbiodegradable).

16. List three health effects from synthetic organics.

    a. ____________________ b. ____________________ c. ____________________

17. A particularly dangerous group of synthetic organic compounds are the ____________________ hydrocarbons.

18. The main feature of halogenated hydrocarbons is that one or more hydrogen atoms have been substituted by an atom of ____________________, ____________________, ____________________, or ____________________.

19. The most common halogenated hydrocarbons are those containing chlorine. Such compounds are referred to as ____________________ hydrocarbons.

20. Examples of chlorinated hydrocarbons include:

    a. ____________________ b. ____________________ c. ____________________

    d. ____________________ e. ____________________

Involvement with Food Chains

21. Bioaccumulation occurs because toxic compounds are

    a. biodegradable (True or False)
    b. nonbiodegradable (True or False)
    c. readily absorbed by the body (True or False)
    d. readily excreted from the body (True or False)

22. Biomagnification demonstrates how each higher trophic level receives and accumulates a (higher, lower, the same) dose than the one before.

23. Mercury caused a biomagnification episode called ________________________ disease.

24. Develop the steps in the food chain that led to the biomagnification episode in Minamata, Japan. Give the correct sequence of events [a to e] that led to the disaster.

Steps in the Food Chain	Events
Step 5	a. mercury in detritus
Step 4	b. biomagnification by fish
Step 3	c. discharge of mercury wastes into river
Step 2	d. Minamata disease in humans
Step 1	e. biomagnification by bacteria

25. Why was biomagnification of mercury in cats not a critical step in causing Minamata disease in humans?

    ____

A History of Mismanagement

26. Prior to the passing of environmental laws in the 1960s and 1970s, it was common practice to dispose of chemical wastes into the ________________ or natural ________________.

27. The public outcry against pollution in the 1960s led the U.S. Congress to pass two major pieces of legislation called:

    a. ____________________________________ Date ________________

    b. ____________________________________ Date ________________

28. In the years since these laws were passed, direct discharges of wastes into natural waterways and the air have (decreased, increased).

Methods of Land Disposal

29. Indicate whether the following statements describe the disposal technique of [a] deep well injection, [b] surface impoundments, or [c] landfills.

    [ ] concentrated wastes put into drums and buried
    [ ] a reverse well

[ ] an open pond without an outlet
[ ] volatile wastes may enter the atmosphere
[ ] waste put into deep rock strata
[ ] involves the removal of leachate

30. Indicate whether the following problems of improper disposal pertain to [a] deep well injection, [b] surface impoundments, [c] landfills, or [d] all three.

[ ] absence of liners or leachate collection systems
[ ] deposition of wastes above or into aquifers
[ ] immersion of wastes into groundwater
[ ] lack of reliable monitoring systems to detect leakage

31. (True or False) Once wastes have seeped into groundwater, there is no practical way to remove them.

32. Small amounts of toxic chemicals in groundwater are extremely hazardous because of their known ability to ________________ and cause serious ________________ ________________.

33. (True or False) When land disposal facilities are properly sited, constructed, and maintained, they guarantee that wastes will never seep or leach into groundwater.

34. Two problems inherent in land disposal is that wastes ________________ at disposal facilities and that they ________________ there.

Scope of the Management Problem

35. The use of designated disposal sites that have no precautions against ground pollution are examples of ________________ land disposal. Give an example. ________________

36. List the three major aspects to the toxic waste problem.

a. ____

b. ____

c. ____

Cleaning Up the Mess

Assuring Safe Drinking Water

37. A federal law that attempts to assure safe drinking water supplies is the ________________________ Date ____________

38. Indicate whether the following are [+] or are not [-] provisions of the Safe Drinking Water Act of 1974.

[ ] setting standards regarding allowable levels of pollution

[ ] monitoring municipal water supplies
[ ] proper location and construction of injection wells
[ ] systematic monitoring of groundwater and private wells
[ ] determination of safe and unsafe levels for all chemicals
[ ] provision of funds to compensate for illness resulting from contaminated wells

39. Identify three emerging issues in the public debate over the Safe Drinking Water Act of 1974.

a. ______________________________

b. ______________________________

c. ______________________________

Groundwater Remediation

40. List the four basic steps in groundwater remediation.

a. ____________________ b. ____________________

c. ____________________ d. ____________________

Superfund for Toxic Sites

41. The Comprehensive Environmental Response, Compensation, and Liability Act of 1980 is commonly known as the ____________________.

42. How is this act funded? ______________________________

43. Funding from the program was about __________ million/year in the early 1980s to above __________ billion/year in 2000.

44. (True or False) All toxic waste sites will be cleaned up in the near future.

45. What is the first step in the types of actions funded by the Superfund?

____

46. What action is taken concerning the worst sites?

____

47. The administrative unit for the Superfund is the ____________________.

48. List four soil cleanup technologies.

a. ____________________ b. ____________________

c. ____________________ d. ____________________

49. Indicate whether the following have been positive [+] or negative [-] attributes of the EPA's administration of the Superfund.

[ ] placement of NPL sites on the Superfund list
[ ] scope of action across all NPL sites in the United States
[ ] development of technologies to clean-up NPL sites
[ ] cost of cleaning up NPL sites
[ ] cost of administering the Superfund

50. How do brownfields demonstrate an example of environmental racism?

____

Management of New Wastes

51. Identify the general roles that the following government entities play in the management of toxic wastes.

a. U.S. Congress: ____

b. EPA: ____

The Clean Air and Water Acts

52. (True or False) The Clean Water Act has stopped nearly all dumping of toxic wastes into natural waterways.

53. What is the purpose of discharge permits?

____

The Resource Conservation and Recovery Act (RCRA)

54. The RCRA legislation required

a. permitting of disposal sites (True or False)
b. pretreatment of wastes destined for landfills (True or False)
c. cradle to grave tracking of all hazardous wastes (True or False)

55. Indicate whether the following are known [+] or anticipated [?] results of the RCRA legislation.

[ ] the closure of many waste disposal facilities
[ ] disposal of wastes in unregulated landfills
[ ] legally permitted discharges
[ ] new methods of midnight dumping
[ ] tougher laws in some states
[ ] shipment of hazardous wastes to other countries
[ ] wholesale movement of some industries to other countries

Reduction of Accidents and Accidental Exposures

56. Match the four potential sources of accidental exposures to hazardous wastes with the safeguards or legislation that have been developed to prevent accidental exposures.

Sources of Exposure	Safeguards
[ ] leaking underground storage tanks	a. EPCRA
[ ] transportation accidents	b. SARA
[ ] workplace accidents	c. TSCA
[ ] new chemicals	d. UST
[ ] worker protection	e. OSHA
[ ] community protection	f. DOT Regs

Looking Toward the Future

57. The EPA estimates of annual costs of pollution control and clean-up programs is more than $__________ billion.

Too Many or Too Few Regulations?

58. There are __________ million tons of toxic chemicals legally discharged into the environment annually.

59. Two groups that are usually exempted from toxic waste regulation compliance are __________________ and ___________________.

60. Why are these groups exempted? ____

Pollution Avoidance for a Sustainable Society

61. List four general approaches in pollution prevention programs.

a. ____

b. ____

c. ____

d. ____

VOCABULARY DRILL					
Directions: Match each term with the list of definitions and examples given in the tables below. Term definitions and examples might be used more than once.					

Term	Definition	Example(s)	Term	Definition	Example(s)
bioaccumulation			midnight dumping		
biomagnification			NPL		
bioremediation			organic chlorides		
brown fields			orphan sites		
ground water remediation			pollution avoidance		
halogenated hydrocarbons			pollution control		
HAZMAT			product life cycle		

VOCABULARY DEFINITIONS
a. a chemical that presents a hazard or a risk
b. cradle-to-grave product history
c. organic compounds in which one or more hydrogen atoms have been replaced by halogens
d. chlorinated hydrocarbons
e. occurs when harmless amounts of toxic waste received over long periods of time reach toxic levels
f. multiplying effect of toxic wastes through the food chain
g. illegal deposition of toxic wastes in any location available under the cover of darkness
h. a technique to clean contaminated groundwater
i. sites representing the most immediate and severe toxic waste threats
j. utilization of oxygen and organisms to clean up toxic waste sites
k. adding filters to prevent environmental pollution
l. changing the production process so that harmful pollutants are not produced
m. hazardous waste sites without a responsible party to do cleanup
n. abandoned industrial facilities where redevelopment is complicated by possible environmental contamination

VOCABULARY EXAMPLES	
a. electric cars	h. Fig. 20-2
b. biodegradable toxic organic compounds	i. military bases
c. catalytic converter	j. Minamata disease
d. Fig. 20-8	k. Fig. 20-11
e. Hg → detritus → bacteria → fish → humans	l. economically disadvantaged communities
f. gasoline, acids, explosives	m. Fig. 20-9
g. CCl_4	

SELF-TEST

Give an example of a hazardous chemical that fits each of the categories listed below.

1. Toxicity: ____________________

2. Reactivity: ____________________

3. Corrosivity: ____________________

4. Ignitability: ____________________

Match the sources of chemicals entering the environment with an example.

	Sources	Examples
5.	[ ] entire products	a. tanker trucks
6.	[ ] fractional products	b. USTs
7.	[ ] waste by-products	c. your discarded shampoo bottle
8.	[ ] household solvents	d. cleaning fluids
9.	[ ] unused portions	e. heavy metals
10.	[ ] leaks	f. paint solvents
11.	[ ] accidental spills	g. fertilizers

12. The two categories of wastes that are particularly toxic are

a. midnight dumping and inadequate landfills
b. heavy metals and synthetic organics
c. deicing salt and waste road oil
d. sewage sludge and transportation spills

13. ____________________ is the term defining organisms' inability to excrete heavy metals.

14. ____________________ is the term defining the multiplying effect of concentrations of heavy metals in the food chain.

Indicate whether the following statements describe the disposal technique of [a] deep well injection, [b] surface impoundments, [c] landfills, or [d] all three.

15. [ ] concentrated wastes put in drums and buried
16. [ ] a reverse well
17. [ ] an open pond without an outlet
18. [ ] volatile wastes may enter the atmosphere
19. [ ] waste put into deep rock strata
20. [ ] involves the removal of leachate
21. [ ] absence of liners or leachate collection systems
22. [ ] deposition of wastes above or into aquifers
23. [ ] immersion of wastes into groundwater
24. [ ] lack of reliable monitoring systems to detect leakage

25. The federal program aimed at identifying and cleaning up existing toxic waste sites is the

a. Resources Conservation and Recovery Act of 1976
b. Clean Water Act of 1974
c. FIFRA
d. Comprehensive Environmental Response, Compensation, and Liability Act of 1980

Which of the following legislative acts were designed to control disposal of chemical (toxic) wastes? (Check all that apply.)

26. [ ] Clean Air Act of 1970
27. [ ] Clean Water Act of 1972
28. [ ] Safe Drinking Water Act of 1974
29. [ ] Comprehensive Environmental Response, Compensation, and Liability Act of 1980
30. [ ] Resource Conservation and Recovery Act of 1976
31. [ ] Toxic Substances Control Act of 1978

Distinguish the difference in pollution avoidance and control.

32. Avoidance: ____

33. Control: ____

ANSWERS TO STUDY GUIDE QUESTIONS

1. toxicology; 2. a, b, c; 3. lower; 4. different; 5. different; 6. d, c, b, a; 7. loss of raw materials, chemical wastes from production process, risks of accidents or spills in manufacturing process, used shampoo rinsed down the drain, containers and packaging thrown into landfill; 8. true; 9. requires industries to report storage and release of toxic chemicals; 10. heavy metals, nonbiodegradable synthetic organics; 11. lead, mercury, arsenic, cadmium, tin, chromium, zinc, copper; 12. metal-workings, batteries, electronics; 13. enzymes, physiological, neurological; 14. plastics, fibers, rubber, paint, solvents, pesticides, wood; 15. nonbiodegradable; 16. mutagenic, teratogenic, carcinogenic; 17. halogenated; 18. chlorine, bromine, fluorine, iodine; 19. chlorinated; 20. plastics, pesticides, solvents, electric insulation, flame retardants; 21. false, true, true, false; 22. higher; 23. Minamata; 24. d, b, e, a, c; 25. cats not in the food chain leading to humans; 26. sewers, waterways; 27. Clean Air Act, 1970; Clean Water Act, 1972; 28. decreased; 29. c, a, b, b, a, c; 30. c, a, d, d; 31. true; 32. bioaccumulation, health hazards; 33. false; 34. arrive, stay; 35. unregulated, Love Canal; 36. assure safe water, regulating the handling and disposal of wastes, looking for future solutions; 37. Safe Drinking Water Act, 1974; 38. +, +, +, -, -, -; 39. cost of protection against a hazard that does not exist, no provision for systematic monitoring of private wells, public acceptance of bottled water as safe in all cases; 40. drilling wells, pumping out contaminated water, purifying water, reinjecting purified water; 41. Superfund; 42. tax on chemical raw materials; 43. 300, 1.4; 44. false; 45. setting priorities (triage); 46. placed on the NPL; 47. EPA; 48. incineration, bioremediation, detergent injection, phytoremediation; 49. +, -, -, -, -; 50. They occur in the most socioeconomic depressed areas of cities; 51. writes the legislation and passes laws, administers programs and enforces laws; 52. false; 53. provide a means of seeing who is discharging what; 54. all true; 55. all +; 56. d, f, e, c, e, a and b; 57. 100; 58. 2; 59. homeowners, farmers; 60. produce < 200 lbs. hazardous mat; 61. better product and/or materials management, find nonhazardous substitute for hazardous material, clean and recyle solvents and lubricants, putting environmentalists and industrialists on the same team

---

ANSWERS TO SELF-TEST

1. pesticides; 2. explosives; 3. acids; 4. gasoline; 5. g; 6. f; 7. e; 8. d; 9. c; 10. b; 11. a; 12. b; 13. bioaccumulation; 14. biomagnification; 15. c; 16. a; 17. b; 18. b; 19. a; 20. c; 21. c; 22. a; 23. d; 24. d; 25. d; 26 - 30. check; 31. blank; 32. does not produce wastes in the first place; 33. filters existing pollution

VOCABULARY DRILL ANSWERS					
Term	**Definition**	**Example(s)**	**Term**	**Definition**	**Example(s)**
bioaccumulation	e	j	midnight dumping	g	d
biomagnification	f	e	NPL	i	i
bioremediation	j	b	organic chlorides	c, d	g
brownfields	n	l	orphan sites	m	m
ground water remediation	h	k	pollution avoidance	l	a
halogenated hydrocarbons	c	g	pollution control	k	c
HAZMAT	a	f	product life cycle	b	h

CHAPTER 21

THE ATMOSPHERE: CLIMATE, CLIMATE CHANGE, AND OZONE DEPLETION

Atmosphere and Weather

Atmospheric Structure

1. Indicate whether the following statements pertain to the [a] troposphere, [b] tropopause, or [c] stratosphere.

   [ ] contains ozone
   [ ] extends up 10 miles
   [ ] caps the troposphere
   [ ] gets colder with altitude
   [ ] temperature increases with altitude

[ ] well-mixed vertically
[ ] site and source of our weather
[ ] substances that enter remain there a long time
[ ] pollutants can reach the top within a few days

Weather

2. What happens to most of the solar radiation that enters Earth's atmosphere?

____

3. How are trade winds produced?

____

4. How are jet streams produced?

____

5. How are tornadoes produced?

____

Climate

6. The average temperature and precipitation conditions of an area produced different _______________

7. (True or False) Humans are isolated from the effects of climate change on other organisms.

Climates in the Past

8. At this time is Earth in a global (cooling or warming) period?

9. Explain the association between variations in Earth's orbit and changing climates.

____

10. Any explanations of rapid global climatic changes must consider the _______________ and the _______________.

Ocean and Atmosphere

11. What proportion of Earth is covered by oceans? __________

12. Oceans are the major source of:

a. _______________ and b. _______________.

13. What two roles do the oceans play in heating the atmosphere?

a. ______________________ and b. ________________________

14. List the two abiotic factors that make the Conveyor system work.

a. ______________________ and b. ________________________

15. Water masses in the Conveyor system move according to _________________.

16. Characterize the following oceanic waters in the Conveyor system in terms of More [M] or Less [L] density.

[ ] Gulf Stream
[ ] North Atlantic Deep Water
[ ] Indian and Pacific deep currents

17. It takes the Conveyor system about __________ years to complete one cycle.

18. Which oceanic waters transfer enormous quantities of heat toward Europe?

a. Gulf Stream b. North Atlantic Deep c. Pacific deep currents

19. Does the addition of freshwater (increase or decrease) ocean water density.

20. Complete the following sequence of events by choosing the bracketed answers.

Extended global warming → [+ or -] melted polar ice → [+ or -] ocean water density → [northern or southern] shift in the Conveyor system → global [cooling or warming].

Global Climate Change

The Earth as a Greenhouse

21. Light energy is (blocked, absorbed) by greenhouse gases.

22. Light energy is converted to __________________ energy in the form of infrared radiation.

23. When heat energy is trapped in an enclosed space and the temperature rises, this is known as the ______________________ effect.

24. What experimental evidence demonstrates a 60 percent increase of atmospheric CO_2 since the ice ages?

____

25. Which of the following factors contribute to planetary albedo? (Check all that apply.)

[ ] atmospheric CO_2
[ ] low clouds
[ ] high clouds

[ ] sulfate aerosols
[ ] atmospheric volcanic ash

The Carbon Dioxide Story

26. CO_2 levels are now __________ percent higher than before the Industrial Revolution.

27. As carbon dioxide concentrations increase, global temperatures will (increase, decrease).

28. A major sink for atmospheric CO_2 is the __________________.

29. Every kilogram of fossil fuel that is burned results in the production of an additional __________ kilograms of carbon dioxide.

30. Forests are a net (source or sink) of CO_2?

31. A total of __________ billion tons of carbon dioxide is added annually to Earth's atmosphere from burning fossil fuels and tropical rain forests.

Other Greenhouse Gases

32. Identify the sources of four other greenhouse gases.

Other Green House Gases	Sources
a. water vapor	
b. methane	
c. nitrous oxide	
d. CFCs and other halocarbons	

Amount of Warming and Its Probable Effects

33. Identify four variables that affect global temperatures.

a. ____________________ b. ____________________

c. ____________________ d. ____________________

34. What is the purpose of general circulation models?

____

35. If concentrations of greenhouse gases were to double, Earth would warm up between __________ and __________ °C.

36. Earth's temperatures were only __________ °C cooler during the ice age.

37. Two predicted major impacts of global warming are

a. ______________________ b. ______________________

38. Indicate whether the following are [+] or are not [-] probable effects of global warming.

[ ] melting of polar ice caps
[ ] flooding of coastal areas
[ ] massive migrations of people inland
[ ] alteration of rainfall patterns
[ ] deserts becoming farmland and farmland becoming deserts
[ ] significant losses in crop yields

Coping with Global Warming

39. List three sources of evidence that global warming might be here.

a. ______________________ b. ______________________

c. ______________________

40. Which of the following are strategies to reduce carbon dioxide emissions and the effects of global warming? (Check all that apply.)

[ ] limit the use of fossil fuels in industry and transportation
[ ] adopt a wait-and-see attitude
[ ] develop and adopt alternative energy sources
[ ] encourage vast tree planting programs
[ ] examine other possible causes of global warming
[ ] make and enforce energy conservation rules
[ ] rely on the government to develop the needed strategies
[ ] adopt the Precautionary Principle

41. Why did the Framework Convention on Climate Change (FCCC) fail?

____

42. After adopting the FCCC, CO_2 emissions of member nations (increased or decreased)?

43. The IPCC calculates that it would take emission reductions of at least __________ percent worldwide to stabilize greenhouse gas concentrations at today's levels. Is it highly (likely or unlikely) that this level of reduction will be achieved.

Depletion of the Ozone Layer

Radiation and Importance of the Shield

44. Ultraviolet light is responsible for thousands of cases of __________________ cancer per year in the United States.

45. ___________________ in the stratosphere prevents ultraviolet light from entering Earth's atmosphere.

46. The presence of this chemical barrier to ultraviolet light is known as the _________________ shield.

47. (True or False) The ozone in the stratosphere and the ozone that is a serious pollutant on the Earth's surface are the same ozone in terms of how they are produced.

Formation and Breakdown of the Shield

48. Complete the following reactions that form ozone.

Reaction 1: ________ light + O_2 $\rightarrow$ __________ + __________

Reaction 2: free ________ + O_2 $\rightarrow$ ________

Reaction 3: free _________ + O_3 $\rightarrow$ __________ + __________

Reaction 4: ________ light + O_3 $\rightarrow$ __________ + ________

49. Indicate whether O_3 concentrations are high [+] or low [-] given the below conditions.

[ ] at the equator [ ] northern latitudes [ ] winter [ ] summer

50. Identify four sources of halogens or chlorofluorocarbons (CFCs).

a. ______________________________ b. ______________________________

c. ______________________________ d. ______________________________

51. A ____________________ is a chemical that promotes a chemical reaction without itself being used up by the reaction.

52. Complete the following reactions that destroy ozone.

Reaction 5: $CFCl_3$ + _________ light $\rightarrow$ _________ + $CFCl_2$

Reaction 6: Cl + _________ $\rightarrow$ ________ + O_2

Reaction 7: ClO + _______ $\rightarrow$ 2 _______ + O_2

53. The source of chlorine entering the stratosphere is ______________________.

54. Which of the above reactions releases Cl from CFCs (5, 6, 7)?

55. Which of the above reactions generates more Cl (5, 6, 7)?

56. Chlorine is a catalyst that destroys the product of reaction (1, 2, 3, 4)?

57. Less ozone in the stratosphere will cause (more or less) UV light penetration.

58. (True or False) CFCs will eventually be flushed out of the atmosphere.

59. (True or False) CFCs are water insoluble.

60. A gaping hole in the ozone shield was discovered over the ___________________ pole.

61. This hole represented a ___________ percent reduction in ozone levels.

62. The ozone hole:

a. persists throughout the year (True or False)
b. reappears every year over the South Pole (True or False)
c. is becoming larger at each appearance (True or False)
d. may cause the destruction of marine phytoplankton (True or False)
e. could be become more destructive if located over the North Pole (True or False)

63. Where is the ozone shield the thinnest? __________________________

64. What is the predicted public health impact in this area?

____

65. (True or False) Volcanic eruptions inject significant amounts of Cl into the stratosphere.

66. (True or False) The Cl in the stratosphere from volcanic eruptions equals that of anthropogenic sources.

67. Indicate whether the following are [+] or are not [-] trends in the continuing saga of ozone depletion.

a. [ ] high ClO over both poles
b. [ ] ozone holes over both poles
c. [ ] ground-level UVB increases
d. [ ] skin cancer increases

68. Explain why ozone loss is expected to peak in 2001.

____

Coming to Grips with Ozone Depletion

69. (True or False) Several nations and industries are scaling back CFCs production.

70. What was the purpose of the Montreal Protocol?

____

71. (True or False) There are CFC substitutes for industrial use?

72. U.S. chemical companies must halt all CFC production by December 31, 19_____.

73. What was the purpose of Title IV of the Clean Air Act of 1990?

____

VOCABULARY DRILL

Directions: Match each term with the list of definitions and examples given in the tables below. Term definitions and examples might be used more than once.

Term	Definition	Example(s)	Term	Definition	Example(s)
anthropogenic			fronts		
catalyst			greenhouse gases		
chlorine cycle			meteorology		
chlorine reservoirs			monsoons		
chlorofluorocarbon			ozone shield		
climate			planetary albedo		
convection currents			weather		

VOCABULARY DEFINITIONS
a. coming from human activities
b. day-to-day variations in temperature, air pressure, wind, humidity and precipitation
c. scientific study of weather and climate
d. result of long-term weather patterns in a region
e. vertical air currents resulting from rising warmer air
f. boundaries where air masses of different temperatures and pressures meet
g. atmospheric gases that play a role analogous to the glass in a greenhouse
h. a natural reflection of light that contributes to overall cooling
i ozone in the stratosphere
j. organic molecules in which Cl and Fl replace some of the hydrogens
k. continual regeneration of Cl as it interacts with ozone
l. promotes a chemical reaction without itself being used up in the reaction

VOCABULARY DEFINITIONS
m. temporary reactions that hold Cl out of the chlorine cycle
n. seasonal air flows that represent a reversal in previous wind patterns

VOCABULARY EXAMPLES	
a. biomes	g. halogenated hydrocarbons
b. low clouds	h. tornadoes
c. CFCs, CO_2, CH_4	i. Indian subcontinent summer weather
d. Cl	j. weather predictions
e. $ClO + ClO \rightarrow 2\ Cl + O_2$	k. NO_2, CH_4
f. Fig. 21-4	l. example given in definition

SELF-TEST

1. The site and source of our weather resides in the

   a. troposphere b. stratosphere c. mesosphere d. thermosphere

2. That part of the atmospheric structure that blocks UV radiation is the

   a. troposphere b. stratosphere c. mesosphere d. thermosphere

3. The Earth's weather engine consists of

   a. atmosphere b. ocean c. solar radiation d. all of these

4. When air masses of different temperature and pressure meet, the weather could be a

   a. tornado b. monsoon c. typhoon d. any one of these

5. Temperature and moisture are the primary predictors of:
   a. climate b. biomes c. weather d. a and b

Identify two ways that oceans play a dominant role in determining our environment.

6. ______________________ and 7. ______________________

Identify the two major components of the **Conveyor** system.

8. ______________________ and 9. ______________________

Complete the following sequence of events by choosing the bracketed choices.

Extended global warming → [10. + or -] melted polar ice → [11. + or -] ocean water density → [12. northern or southern] shift in the Conveyor system → global [13. cooling or warming].

Identify the source and corresponding pollutant that cause each of the following environmental conditions.

Condition	Sources	Pollutant
Greenhouse Gases	14.	15.
	16.	17.
	18.	19.
	20.	21.
Depletion O_3 Shield	22.	23.
	24.	25.

26. Define global warming and apply the term to climatic change: ____________________

____

Indicate whether the following environmental conditions are or might be the result of or result in [A] planetary albedo, [B] greenhouse effect, or [C] ozone depletion.

27. [ ] increased concentrations of UVB light hitting the Earth's surface
28. [ ] melting of polar ice caps
29. [ ] deserts becoming farmland and farmland becoming deserts
30. [ ] global cooling
31. [ ] higher incidence of skin cancer in humans
32. [ ] volcanic eruptions
33. [ ] flooding of coastal areas.

Identify whether the following pollution abatement procedures are [A] getting at the root cause or [B] attacking the symptoms.

34. _____ decrease fertilizer use 35. _____ banning aerosol sprays 36. _____ using solar energy

37. _____ recapture CFCs 38. _____ increase fuel efficiency 39. _____ increase fuel costs

40. The ozone shield in the stratosphere protects Earth's biota from the damaging effects of

a. CFCs b. UV light c. global warming d. planetary albedo

ANSWERS TO STUDY GUIDE QUESTIONS

1. c, a, b, a, c, a, a, c, a; 2. absorbed by atmosphere, oceans, and land; 3. upward movements of air called convection currents; 4. Earth's rotation and air pressure gradients; 5. convergence of air masses of different temperatures and pressures; 6. climates; 7. false; 8. warming; 9. changes in Earth's orbit cause changes in the distribution of solar radiation over different continents and latitudes, thus changing temperature and precipitation; 10. atmosphere, ocean; 11. 2/3; 12. water for hydrological cycle, heat to the atmosphere; 13. stores and conveys heat; 14. temperature and salinity; 15. density; 16. less, more, less; 17. 1000; 18. a; 19. decrease; 20. +, -, southern, cooling; 21. absorbed; 22. heat; 23. greenhouse; 24. CO_2 in gas bubbles trapped in glacier ice; 25. blank, check, blank, check, check; 26. 35; 27. increase; 28. ocean; 29. 3; 30. source; 31. 24; 32. water cycle, animal husbandry, chemical fertilizers, refrigerants; 33. cloud cover, solar intensity, volcanic activity, sulfate aerosols; 34. project future global climate; 35. 1 to 4.5; 36. 5; 37. regional climatic changes, rise in sea level; 38. all +; 39. nine out of 14 hottest years, wide-scale recession of glaciers, sea level is rising; 40. check, blank, check, check, blank, check, blank, check; 41. relied on a voluntary approach; 42. increased; 43. 60, unlikely; 44. skin; 45. ozone; 46. ozone; 47. false; 48. UVB, O, O; O, O_3; O, O_2, O_2; UVB, O, O; 49. +, -, -, +; 50. refrigerants, plastic foams, electronics cleaners, aerosol cans; 51. catalyst; 52. UV, Cl; O_3, ClO; ClO, 2Cl; 53. CFCs; 54. 5; 55. 7; 56. 2; 57. more; 58. false; 59. true; 60. south; 61. 50; 62. false, the rest are all true; 63. Queensland, Australia; 64. 3/4 Australians will develop skin cancer; 65. false; 66. false; 67. +, -, +, +; 68. anthropogenic sources of Cl and Br in stratosphere will start to decline; 69. true; 70. an agreement by 68 nations to scale back CFC production 50 percent by 2000; 71. true; 72. 1995; 73. restricts the production, use, emissions, and disposal of a family of chemicals identified as ozone depleting

---

ANSWERS TO SELF-TEST

1. a; 2. b; 3. d; 4. d; 5. d; 6.convey heat; 7. store heat; 8. salinity; 9. density; 10. +; 11. -; 12. southern; 13. cooling; 14. coal; 15. carbon dioxide; 16. cattle; 17. methane gas; 18. refrigerants; 19. CFCs; 20. gasoline; 21. nitrous oxides; 22. refrigerants; 23. CFCs; 24. plastic foams; 25. electronics; 26. global warming is a process of gradual increase in Earth's temperature, which may cause a southern shift in the Gulf Stream, leading to another ice age; 27. C; 28. B; 29. B; 30. A; 31. C; 32. A; 33. B; 34. A; 35. A; 36. A; 37. A; 38. B; 39. B; 40. B.

VOCABULARY DRILL ANSWERS

Term	Definition	Example(s)	Term	Definition	Example(s)
anthropogenic	a	c	fronts	f	h
catalyst	l	d	greenhouse gases	g	c
chlorine cycle	k	e	meterology	c	j
chlorine reservoirs	m	k	monsoons	n	i
chlorofluorocarbon	j	g	ozone shield	i	l
climate	d	a	planetary albedo	h	b
convection currents	e	f	weather	b	h

CHAPTER 22

ATMOSPHERIC POLLUTION

Air Pollution Essentials

Pollutants and Atmospheric Cleansing

1. List two processes that hold natural pollutants below the toxic level.

   a. ______________________________

   b. ______________________________

2. List the three factors that determine air pollution levels.

   a. ____________________ b. ____________________

   c. ____________________

The Appearance of Smogs

3. The discovery of ______________ began human inputs of air pollutants.

4. In Hard Times, Charles Dickens referred to ________________ smog.

5. Thousands of cars venting unprotected exhaust + sunlight + a mountain topography = ________________ smog.

6. Normally temperature (increases, decreases) as elevation increases.

7. The warmer surface air ________________, which tends to carry pollutants away.

8. Surface air cannot rise when a layer of (warmer, cooler) air overlies cooler air, because this creates a situation that is referred to as a ______________________ inversion.

9. What is the highest recorded number of human deaths due to an air pollution disaster? __________

10. Air pollution may

    a. adversely affect human health (True or False)
    b. cause damage to crops and forests (True or False)
    c. increase the rate of corrosion and deterioration of materials (True or False)

11. (True or False) Human illnesses and damage to crops and forests from air pollution has become increasingly commonplace.

Major Air Pollutants and Their Impact

Major Pollutants

12. List the eight pollutants or pollutant categories that are the most widespread and serious.

    a. ________________________ b. ______________________ c. ______________________

    d. ________________________ e. ______________________ f. ______________________

    g. ________________________ h. ______________________

13. In large measure, all of the above pollutants are direct or indirect products of fossil fuel ____________________.

Adverse Effects of Air Pollution on Humans, Plants, and Materials

14. Air pollution

    a. is the result of one chemical mixed with the normal constituents of air (True or False)
    b. varies in time and place (True or False)
    c. varies with environmental conditions (True or False)

15. It is (easy or difficult) to determine the role of particular pollutants in causing an observed result.

16. Identify the three categories of air pollutant impacts on humans.

    a. ____________________ b. ____________________ c. ____________________

17. Serious adverse effects of air pollution on human health are seen

    a. mainly among smokers b. mainly among nonsmokers c. equally among both groups

18. (True or False) Smoking by itself greatly increases the risk of serious disease.

19. Learning disabilities in children and high blood pressure in adults are correlated with high levels of ____________________ in the blood.

20. The source of widespread lead contamination was

    a. people who smoke b. gasoline c. high ozone levels d. pesticide use

21. Air pollution

    a. may destroy vegetation (True or False)
    b. may seriously retard the growth of crops and forests (True or False)
    c. may cause forests to become vulnerable to insect pests and disease (True or False)

22. Indicate whether the following effects of air pollutants have [+] or have not [-] been demonstrated.

[ ] which pollutants cause the damage
[ ] the sensitivity level of plants to gaseous air pollutants
[ ] the degree to which air pollutants retard plant growth and development
[ ] the degree to which air pollutants reduce crop yields and profits

23. The most serious pollutant affecting agriculture and natural areas is ______________.

24. (True or False) Crops and wild plants react in the same manner to air pollutants.

25. (True or False) Just a small increase in concentration or duration of exposure may push some plants beyond their ability to cope with an air pollutant.

26. Indicate whether the following conditions could [+] or could not [-] be an effect of air pollution.

[ ] gray and dingy walls and windows
[ ] deteriorated paint and fabrics
[ ] corrosion of metals
[ ] a bright, clear, blue sky
[ ] increased real estate values
[ ] decreased real estate values

Sources of Pollutants

27. Air pollutants are direct or indirect by-products from burning ________________, ________________ and ____________________.

28. Oxidation of coal, gasoline, and refuse by burning (is, is not) usually complete.

Primary Pollutants

29. Indicate which of the following are [+] or are not [-] primary pollutants from combustion.

[ ] sulfuric acid
[ ] hydrocarbon emissions
[ ] ozone
[ ] nitric oxide
[ ] sulfur dioxide
[ ] carbon monoxide
[ ] PANs
[ ] particulates
[ ] nitrogen oxides
[ ] lead
[ ] photochemical oxidants
[ ] volatile organic compounds (VOCs)

Secondary Pollutants

30. Indicate which of the following are [+] or are not [-] secondary pollutants from combustion.

[ ] sulfuric acid
[ ] hydrocarbon emissions
[ ] ozone
[ ] nitric oxide
[ ] sulfur dioxide
[ ] carbon monoxide
[ ] PANs
[ ] particulates
[ ] nitrogen oxides
[ ] lead
[ ] photochemical oxidants
[ ] volatile organic compounds (VOC's)

31. Match the major source(s) of the following pollutants (see Fig. 22-13).

Pollutant	Source(s)
sulfur oxide	a. fuel combustion
nitrogen oxides	b. transportation
volatile organic compounds	c. industrial processes
PM-10 particulate matter	d. miscellaneous sources
carbon monoxide	

Acid Deposition

32. Industrialized regions of the world are regularly experiencing precipitation that is from 10 to ____________ times more acid than normal.

33. List four forms of acid precipitation.

a. ________________________ b. ________________________

c. ________________________ d. ________________________

Acids and Bases

34. Indicate whether the following are properties of [a] acids, [b] bases, or [c] water.

[ ] any chemical that will release OH^- ions
[ ] any chemical that will release H^+ ions
[ ] the combination of one OH^- and 2 H^+ ions
[ ] the balance or neutral point between acidic and basic solutions

35. pH is a measurement of (hydrogen, hydroxyl) ions in solution.

36. On the pH scale, the number ____________ represents the pH of pure water or neutral solution

37. On the pH scale, numbers decreasing from 7 (e.g., 6, 5, 4, 3, 2, 1) represent increasing (aAcidity, aAlkalinity).

38. Each unit on the pH scale represents a factor of (1, 10, 100, 1000) in the acid concentration of a solution.

39. A solution of pH 5 is (1, 10, 100, 1000) times more acid than a solution of pH 6.

40. A solution of pH 4 is (1, 10, 100, 1000) times more acid than a solution of pH 6.

Extent and Potency of Acid Deposition

41. Acid precipitation is defined as any precipitation with a pH of ___________ or less.

42. Give the pH values for the following examples.

a. rainfall in the eastern United States from ___________ to ___________ (see Fig. 22-18)
b. mountain forests east of Los Angeles: __________

Sources of Acid Deposition

43. About two thirds of the acid in acid precipitation is ____________________ acid and one third is ____________________ acid.

44. It is well known that sulfur and nitrogen are found in fossil ________________.

45. Indicate whether the following statements pertain to anthropogenic [a] or natural [n] oxide emissions.

[ ] produced by volcanic or lightning activity
[ ] produced by human activities
[ ] concentrated in industrial regions
[ ] spread out over the globe

46. In the eastern United States, the source of more than 50 percent of acid deposition are ________________ burning power plants.

47. Power plants attempt to alleviate sulfur dioxide emission at ground level by building (shorter, taller) stacks that ultimately disperse the pollutant (closer, further) from the source.

Effects of Acid Deposition

48. pH controls all aspects of ________________, ________________, and ________________.

49. Indicate whether the following conditions are [+] or are not [-] known effects of acid pH in aquatic ecosystems.

[ ] alteration of plant and animal reproduction
[ ] introduction of toxic elements (e.g., aluminum) from soil leachate

[ ] shift from eutrophic to seemingly oligotrophic conditions
[ ] totally barren and lifeless aquatic ecosystems
[ ] alterations in the food chains

50. Suppose two regions receive roughly equal amounts of acid precipitation. How is it possible for one region to have acidified lakes while the other region does not?

____

51. (True or False) Lakes can maintain their buffering capacity indefinitely.

52. Hydrogen ion concentrations in solution will (increase, decrease) in the presence of a buffer.

53. A mineral that is an important buffer in nature is ____________________, known chemically as calcium carbonate.

54. The amount of acid that a given amount of limestone can neutralize is known as its ________________ capacity.

55. When an ecosystem loses its buffering capacity

a. additional hydrogen ions will remain in solution (True or False)
b. the buffer is used up (True or False)
c. the ecosystem will become acidified (True or False)

56. When the buffering capacity is exhausted, there is a (rapid, slow) drop in pH.

57. Indicate whether the following tree species are tolerant [T] or vulnerable [V] to acid rain.

[ ] red spruce [ ] balsam fir [ ] sugar maple

58. Indicate whether the following are direct [d] or indirect [i] effects of acid precipitation on forest ecosystems. (This is not covered in the text directly, but think about it!)

[ ] leaching of nutrients
[ ] breakdown and release of aluminum into solution
[ ] rapid changes in soil chemistry
[ ] reduced growth and diebacks of plant and animal populations
[ ] increased plant vulnerability to natural enemies and drought
[ ] increased soil erosion
[ ] increased flooding
[ ] increased sedimentation of waterways

59. Indicate whether the following effects of acid precipitation on humans is related to [a] artifacts, [b] health, or [c] aesthetics.

[ ] mobilization of lead and other toxic elements
[ ] corrosion of limestone and marble
[ ] deterioration of lakes and forests

Bringing Air Pollution Under Control

60. The law that mandated setting standards for air pollutants is the _______________.

61. The Clean Air Act mandated setting standards for

    a. all air pollutants (True or False)
    b. five pollutants recognized as most widespread and objectionable (True or False)

62. Explain the difference between ambient and primary standards.

    ____

    ____

63. List the five pollutants that are most widespread and objectionable.

    a. ____________________ b. ____________________ c. ____________________

    d. ____________________ e. ____________________

Control Strategies

64. Explain the command-and-control approach.

    ____

    ____

65. Indicate the level, high [H] or low [L], of success of the command-and-control approach for the following air pollutants.

    a. [ ] lead b. [ ] ozone c. [ ] particulate matter d. [ ] SO_2

66. Identify two devices that some industries have installed to reduce particulates.

    a. ____________________ b. ____________________

67. (True or False) Filters and electrostatic precipitators remove all toxic substances.

68. List four contemporary sources of particulates.

    a. ____________________ b. ____________________

    c. ____________________ d. ____________________

69. How does the Clean Air Act of 1990 address the 83 regions of the Unites States that have failed to attain air particulates standards?

    ____

70. A new car today emits ________ percent less pollutants than pre-1970 cars.

71. Explain how a catalytic converter works.

____

____

72. List the three major considerations of the Clean Air Act of 1990 concerning motor vehicle pollutants.

a. ____

b. ____

c. ____

73. Identify a point and area source of VOC emissions.

Point: ______________________ Area: ______________________

74. List two VOC control strategies in the Clean Air Act of 1990.

a. ____

b. ____

75. The estimated total toxic chemicals emitted into the air in the United States is around __________ million tons annually.

76. At least __________ toxic pollutants have been identified.

Coping with Acid Precipitation

77. How much of a reduction in acid-causing emissions is required to prevent further acidification of the environment? __________ percent

78. Identify the major problems (a to e) associated with the adoption of any of the following ways to reduce acid-forming emissions.

[a.] expense
[b.] major reconstruction of industrial facility
[c.] societal acceptance
[d.] requires major shift in U.S. energy policy
[e.] still have the problem of where to put the waste

[ ] fuel switching
[ ] coal washing
[ ] fluidized bed combustion

[ ] scrubbers
[ ] alternative power plants
[ ] reductions in electricity consumption

79. Indicate whether the following strategies for reducing acid-producing emissions are examples of [a] fuel switching, [b] coal washing, [c] fluidized bed combustion, [d] scrubbers, or [e] alternative power plants.

[ ] exhausting fumes through a spray of water containing lime
[ ] combustion of coal in a mixture of sand and lime
[ ] building more nuclear power plants
[ ] pulverize and chemically wash coal
[ ] using low-sulfur coals

80. The circumstantial evidence linking power plant emissions to acid deposition is (insufficient, overwhelming).

81. Utility companies have (more, less) governmental support than does the public on the acid precipitation issue.

82. The federal government's response to the overwhelming evidence on the cause-and-effect relationship of acid deposition was more ________________.

83. What made Title IV a unique addition to The Clean Air Act?

____

84. List the five provisions of Title IV.

a. ____

b. ____

c. ____

d. ____

e. ____

85. Which of the following have been industry's response to Title IV? (Check all that apply.)

[ ] fuel switching
[ ] coal washing
[ ] scrubbers
[ ] emissions allowance trading
[ ] using low-sulfur coals

Taking Stock

86. Currently, pollution control costs are estimated at $ ________ billion per year.

87. (True or False) Pollution control is now a major industry.

Future Directions

88. List four future improvements that can be made to reduce air pollution.

a. ______________________________ b. ______________________________

c. ______________________________ d. ______________________________

VOCABULARY DRILL

Directions: Match each term with the list of definitions and examples given in the tables below. Term definitions and examples might be used more than once.

Term	Definition	Example(s)	Term	Definition	Example(s)
acid deposition			catalytic converter		
acid precipitation			chronic		
acids			industrial smog		
acute			pH		
air pollutants			photochemical smog		
ambient standards			point sources		
area sources			primary pollutants		
artifacts			primary standard		
bases			secondary pollutants		
buffer			temperature inversion		
carcinogenic			threshold level		

VOCABULARY DEFINITIONS
a. precipitation that is more acidic than usual
b. concentration of H ions in solution
c. substances in the atmosphere that have harmful effects
d. pollutant level below which no ill effects are observed
e. an irritating grayish mixture of soot, sulfurous compounds and water vapor
f. air pollutant resulting from sunlight interacting with primary sources
g. effect of certain weather conditions on smog levels
h. air pollutant levels that need to be achieved to protect environmental and human health
i. gradual deterioration of a variety of physiological functions over a period of years
j. life-threatening reactions within a period of hours or days
k. cancer-causing
l. direct products of combustion and evaporation
m. products from reactions between primary pollutants and atmospheric conditions
n. the highest level of pollutant that can be tolerated by humans without noticeable ill effects.
o. a device that removes toxic substances from internal combustion engines
p. direct pollution sources
q. diffuse pollution sources
r. any precipitation that is 5.5 or less
s. substance that has a large capacity to absorb H ions and hold pH relatively constant
t. human-made objects
u. H ion > OH ion concentration
v. H ion < OH ion concentration

VOCABULARY EXAMPLES	
a. Fig. 22-12	k. household products
b. bronchitis, fibrosis of the lungs	l. industry
c. Fig. 22-22	m. Table 22-2
d. death	n. PM-10 for particulates
e. Fig. 22-5	o. 0 | 7 | 14
f. Fig. 20-1 (earlier chapter)	p. limestone
g. lungs	q. monuments
h. Fig. 22-6	r. pH > 7
i. Fig. 22-3	s. pH < 7
j. Hard Times by Charles Dickens	t. H_2SO_4

SELF-TEST

1. Which of the following statements is not accurate concerning the air pollution problem?

   a. Organisms have the capacity to deal with certain amounts and levels of pollution.
   b. There is a threshold pollution level that is tolerable.
   c. Air pollution has been with human society since the discovery of fire.
   d. The solution to pollution is dilution.

2. Unprotected car exhaust + sunlight + a mountain topography =

   a. a temperature inversion b. photochemical smog c. an air pollution episode d. acid rain

3. The condition of a warm air mass overlying a cool air mass is referred to as

   a. a temperature inversion b. photochemical smog c. an air pollution episode d. acid rain

4. Air pollution may

   a. adversely affect human health
   b. cause damage to crops and forests
   c. increase the rate of corrosion and deterioration of materials
   d. cause all of the above effects

5. Any chemical that releases H^+ ions is

   a. an acid b. a base c. water d. all of these

6. Measurement of pH is actually a measure (hydrogen, hydroxyl) ions.

7. A solution of pH 5 (1, 10, 100, 1000) times more acid than a solution of pH 6.

8. About two thirds of the acid in acid precipitation is

a. chromic b. nitric c. sulfuric d. carboxylic acid

9. It is well known that the source(s) of acid precipitation is (are)

a. sulfur dioxide in coal
b. nitrogen oxides in gasoline
c. coal fired power plants
d. all of these

Identify four direct or indirect effects of air pollutants on pine trees and the forest ecosystem in general.

10. ______________________________

11. ______________________________

12. ______________________________

13. ______________________________

14. A major problem with the Clean Air Act of 1970 and its 1977 and 1990 amendments is

a. identifying pollutants
b. demonstrating cause and effect relationships
c. determining the source of pollutants
d. developing and implementing controls

15. Indicate whether the following air pollutants are Primary [P] or Secondary [S].

[ ] particulates
[ ] hydrocarbon emissions
[ ] carbon monoxide
[ ] nitric oxide
[ ] sulfur dioxide
[ ] ozone
[ ] PANs
[ ] VOCs
[ ] lead
[ ] sulfuric acid
[ ] nitric acid

16. A major change in the Clean Air Act of 1990 was:

a. identifying criteria pollutants
b. imposition of sanctions on polluters
c. allowing polluters to choose the most cost-effective means of pollution control
d. all of the above

17. Air pollution is known to have adverse effects on

a. human health b. agricultural crops c. forests d. all of these

18. List three effects of air pollution on human health.

a. ______________________________ b. ______________________________

c. ______________________________

19. Indicate whether the strategies described below were designed to control emissions of a. particulates, b. pollutants from motor vehicles, c. sulfur dioxide and acids, d. ozone, or e. air toxics.

[ ] phasing out the use of leaded gasoline
[ ] no visible emissions
[ ] putting catalytic converters on car exhaust systems
[ ] installation of filters and electrostatic precipitators at industrial sites
[ ] tall smoke stacks
[ ] withholding federal highway funds

ANSWERS TO STUDY GUIDE QUESTIONS

1. disperse and dilute in atmosphere, oxidation by hydroxyl ions; 2. amount, space, mechanisms of removal; 3. fire; 4. industrial; 5. photochemical; 6. decreases; 7. rises; 8. warmer, thermal; 9. 4,000; 10. all true; 11. true; 12. suspended particulate matter, volatile organic compounds, carbon monoxide, nitrogen oxides, sulfur oxides, heavy metals, ozone, air toxics; 13. combustion; 14. all true; 15. difficult; 16. chronic, acute, carcinogenic; 17. a; 18. true; 19. lead; 20. b; 21. all true; 22. all +; 23. ozone; 24. false; 25. true; 26. +, +, +, -, -, +; 27. coal, gasoline, refuse; 28. is not; 29. -, +, -, +, +, +, -, +, -, +, -, +; 30. the reverse of #29; 31. a; a, b; b, c; a, b, c; b; 32. 1000; 33. rain, fog, mist, snow; 34. b, a, c, c; 35. hydrogen; 36. 7; 37. acidity; 38. 10; 39. 10; 40. 100; 41. 5.5; 42. 4 to 3, 2.8; 43. sulfuric, nitric; 44. fuels; 45. n, a, a, n; 46. coal; 47. taller, further; 48. enzymes, hormones, proteins; 49. +, +, -, +, +; 50. soil has a natural buffer; 51. false; 52. decrease; 53. limestone; 54. buffering; 55. all true; 56. rapid; 57. V, T, T; 58. d, d, d, d, d, i, i, i; 59. b, a, c; 60. Clean Air Act; 61. false, true; 62. levels that need to be achieved to protect environmental and human health, level of human tolerance without suffering noticeable health effects; 63. particulates, SO_2, CO, NO, O_3; 64. regulate air pollution so criteria pollutants remain below primary standard level; 65. H, L, L, L; 66. filters, electrostatic precipitators; 67. false; 68. steel mills, power plants, smelters, construction sites; 69. region must submit plans based on reasonably available control strategies (RACT); 70. 75; 71. platinum coated beads oxidize VOC to CO_2 and H_2O, oxide CO to CO2, and reduces some NO; 72. tighten emission standards, develop cleaner-burning fuel, drive less; 73. industry, household products; 74. prohibition of some consumer goods, withholding federal highway funds; 75. 4.3; 76. 188; 77. 50; 78. d, a, e, a, b, c; 79. d, c, e, b, a; 80. overwhelming; 81. more; 82. study; 83. addressed acid rain and CFCs specifically; 84. reductions in SO_2 levels, methods of implementing reductions, emissions allowances and trading, emissions purchasing, reductions in NO; 85. check, blank, check, check, check; 86. 150; 87. true; 88. increase fuel efficiency, emission-free vehicles, improving mass transit systems, reduce commuting distances

---

ANSWERS TO SELF-TEST

1. d; 2. b; 3. a; 4. d; 5. a; 6. hydrogen; 7. 10; 8. c; 9. d; 10-13. see study guide question 65; 14. d; 15. P, P, P, P, P, S, S, P, P, S, S; 16. d; 17. d; 18. chronic, acute, carcinogenic; 19. b, a, b, a, c, d

VOCABULARY DRILL ANSWERS					
Term	**Definition**	**Example(s)**	**Term**	**Definition**	**Example(s)**
acid deposition	a	p	catalytic converter	o	c
acid precipitation	r	p	chronic	i	b
acids	u	s, t	industrial smog	e	j
acute	j	d	pH	b	o
air pollutants	c	h, i , j	photochemical smog	f	i
ambient standards	h	n	point sources	p	l
area sources	q	k	primary pollutants	l	a
artifacts	t	q	primary standard	n	m
bases	v	r	secondary pollutants	m	a
buffer	s	p	temperature inversion	g	e
carcinogenic	k	g	threshold level	d	f

CHAPTER 23

ECONOMICS, PUBLIC POLICY, AND THE ENVIRONMENT

1. Indicate whether the following statements are conditions or characteristics of [a] Executive Order 12291, [b] Executive Order 12866, or [c] both Executive Orders.

   [ ] contained the process of public and inter-agency review of proposed regulatory rules
   [ ] requires cost-benefit analysis
   [ ] Qualifying regulations must impose a cost of at least $100 million.
   [ ] Qualifying regulations must have a significant adverse effect on U.S. firms to compete with foreign firms.
   [ ] requires consideration of how proposed action reduces risks to public health

2. (True or False) Cost-benefit analysis is applied in the context of what is right to do.

Economics and Public Policy

The Need for Environmental Public Policy

3. Identify the two areas of emphasis in promoting the common good with environmental public policies.

   a. ______________________________

   b. ______________________________

4. What is the message portrayed in Table 23-1 concerning the need for environmental public policy?

   ____

Relationships Between Economic Development and the Environment

5. Examine Fig. 23-1 and determine whether the relationships between income and environmental indicators are [D] direct, [I] inverse, or both [B].

   [ ] population without safe water
   [ ] population without adequate sanitation
   [ ] urban concentrations of SO_2
   [ ] municipal wastes per capita
   [ ] CO_2 emissions per capita

Economic Systems

6. Identify which of the characteristics below describe a [a] centrally planned economy, [b] free market economy, or [c] both economic systems.

   [ ] characteristic of socialist countries
   [ ] characteristic of capitalistic countries
   [ ] Factors of production are land, labor, and capital.
   [ ] Ruling class makes all the decisions.

[ ] Equity and efficiency are theoretically achievable.
[ ] North Korea is one of the last holdouts of this economic model.
[ ] The market itself determines what will be exchanged.
[ ] The system operates according to the interplay of supply and demand.
[ ] All "players" have free access to the market.
[ ] easily manipulated by powerful business interests
[ ] only offers access to goods and services
[ ] Developed countries use this economic model.

7. (True or False) No country has a pure form of either economy.

Resources and Wealth of Nations

8. What two factors are indicators of a country's natural capital?

a. ______________________ b. ______________________

The Wealth of Nations

9. List the three major capital components that determine a nation's wealth.

a. ______________________ b. ______________________

c. ______________________

10. Match the capital components on the left with an example on the right.

Capital Components	Examples
Produced assets	a. forests, fisheries, soil, water
Renewable natural capital	b. population characteristics and cultural attributes
Nonrenewable natural capital	c. buildings, machinery, highways
Human capital	d. social and political environment
Social capital	e. oil and mineral deposits

11. The dominant force of wealth for most nations is ______________________.

12. The dominance of the human resources component emphasizes the importance of investing in:

a. ______________________ b. ______________________

c. ______________________

Shortcomings of GNP

13. What is a major shortcoming or omission in the calculation of GNP?

____

14. List two reasons why nations will always underestimate the value of their natural resources.

a. ____

b. ____

Resource Distribution

15. Besides controlled population growth, identify two factors that promote intragenerational equity in countries.

a. ______________________ b. ______________________

16. Meeting the needs of the present without compromising the ability of future generations to meet their needs is called ______________________ equity.

17. Indicate whether the following statements are [+] or are not [-] examples of intergenerational equity.

[ ] Let future generations cope with global climate change from fossil fuel consumption.
[ ] Cut all the trees and sell the timber today.
[ ] Hold on to the trees for a better price some time in the future.
[ ] Give up short-term gains for sustainable long-term harvests.

Pollution and Public Policy

Public Policy Development: The Policy Life Cycle

18. Match the following list of policy life cycle characteristics with the four stages: [a] recognition,

[b] formulation, [c] implementation, or [d] control.

[ ] Real political and economic costs of a policy are exacted.
[ ] low in political weight
[ ] policies broadly supported
[ ] media have popularized the policy
[ ] rapidly increasing political weight
[ ] dissension is high
[ ] Public concern and political weight are declining.
[ ] policies broadly supported
[ ] media coverage is high
[ ] issue not very interesting to media
[ ] The environment is improving.
[ ] Debate about policy options occurs.

19. Using the same stages as listed above, indicate the status of policies that address the following environmental problems (see Fig. 23-8).

[ ] toxic chemicals
[ ] sewage water treatment

[ ] global warming
[ ] ozone depletion
[ ] urban sprawl

Economic Effects of Environmental Public Policy

20. The best environmental policies are those that conform to the ______________________, ______________________, and ______________________ criteria.

21. (True or False) Some policies have no direct monetary costs.

22. (True or False) Some policies have no political costs.

23. The relationship between strict environmental regulations and economic growth is (positive or negative)?

24. Summarize the general conclusions of studies on the impact of environmental policy on the economy.

____

Policy Options: Market or Regulatory?

25. Indicate whether the following statements or policy examples characterize the market [M] or regulatory [R] approaches.

[ ] Clean Air Act of 1990
[ ] the most common approach in environmental public policy
[ ] practically guarantees a certain sustained level of pollution

Benefit-cost Analysis

26. Explain why Executive Order 12291 was issued by President Reagan.

____

27. A comparison of costs and benefits is commonly called a ________________ analysis.

28. Cost-effective projects have benefits that are (greater, less) than the costs.

External and Internal Costs

29. The effect of the business process not included in the usual calculations of profit and loss is called an __________________.

30. Give an example of an:

a. external good ____

b. external bad ____

The Costs of Environmental Regulations

31. (True or False) Most forms of pollution control involve additional expenses.

32. List two ways that pollution control increases costs.

    a. ____

    b. ____

33. (True or False) One-hundred percent pollution control is cost-effective (see Fig. 23-11a).

34. Costs of pollution control are the highest in the (early, later) stages of control.

The Benefits of Environmental Regulation

35. List six major benefits derived from pollution control (see Table 23-3).

    a. ____

    b. ____

    c. ____

    d. ____

    e. ____

    f. ____

36. Indicate whether the following benefits or risks can be assigned a monetary value through [a] real dollar estimates, [b] shadow pricing, or [c] cannot be given a value.

    [ ] health benefits from eliminating air pollution episodes
    [ ] costs of maintaining and replacing materials
    [ ] income from water recreation
    [ ] improving air quality
    [ ] impact on nonhuman environmental components
    [ ] assessing the value of human life

37. Give one method of estimating benefits derived from pollution control.

    ____

Cost Effectiveness

38. At what level of pollution control will the benefits be negligible and not worth the cost?
    __________ percent or below ____________________ level

39. Optimum cost-effectiveness is achieved when the benefit curve is the (greatest, least) distance above the cost curve.

40. (True, False) Equipment, labor, and maintenance costs can be estimated with a fair degree of objectivity.

41. Cost of pollution controls will be (high, low) at the time they are initiated and will tend to (increase, decrease) as time passes.

42. (True or False) Some situations that may appear to be cost-ineffective in the short term may be extremely cost-effective in the long term.

43. Identify two types of pollution that may apply to a "true" response to the previous question.

a. ______________________ b. ______________________

44. (True or False) Those who bear the cost of pollution control and those who benefit from these controls are different groups of people.

45. Give one example of a situation that would support a "true" response to the previous question.

____

Progress

46. Indicate whether the following have [+] or have not [-] been EPA achievements from 1970 to the present.

[ ] reduced in vehicle emissions
[ ] constructed upgraded wastewater treatment facilities
[ ] eliminated ocean dumping
[ ] stopped land disposal of untreated hazardous wastes
[ ] cleaned up 520 hazardous waste sites
[ ] banned production and use of asbestos, DDT, PCBs and leaded gasoline

47. The phase out of leaded gasoline cost the oil industry $ __________ billion but accrued benefits represented $ __________ billion.

Politics, the Public and Public Policy

48. List the three main ways Congress influences environmental policy.

a. ____

b. ____

c. ____

Politics and the Environment

49. Strategies to introduce anti-environmental legislation in the 104th and 105th Congresses included

a. attempts to dismantle many key environmental laws (True or False)
b. attachment of anti-environmental riders to budget legislation (True or False)
c. use of deceptive labeling (True or False)

50. List four strategies used by special interest groups to promote their organizational agendas.

a. ____________________

b. ____________________

c. ____________________

d. ____________________

Citizen Involvement

51. Which types of public involvement in policy development match the [a] recognition, [b] formulation, [c] implementation, and [d] control stages in the policy life cycle.

[ ] grassroots concern about environmental problems
[ ] informing constituencies about the problem
[ ] informing elected officials about level of public support for environmental policy
[ ] electing or defeating political candidates
[ ] public comment on regulations

VOCABULARY DRILL

Directions: Match each term with the list of definitions and examples given in the tables below. Term definitions and examples might be used more than once.

Term	Definition	Example(s)	Term	Definition	Example(s)
benefit-cost analysis			natural capital		
cost-effective			policy life cycle		
external cost			produced assets		
GNP			regulatory approach		
human capital			social capital		
market approach			special interest groups		

VOCABULARY DEFINITIONS
a. recognition, formulation, implementation, control
b. set prices on pollution and resource use
c. standards are set and technologies prescribed
d. compares the estimated costs of project with the benefits that will be achieved
e. stock of tangible goods, e.g., buildings, machinery, highways, etc.
f. economic justification for proceeding with a given project
g. an effect of the business process not included in calculations of profit and loss
h. goods and services supplied by natural ecosystems
i. physical, psychological, and cultural attributes of the population
j. social and political environment of a human society
k. sum of all goods and services produced in a country
l. organizations that promote their personal agendas in the policy and decision-making process

VOCABULARY EXAMPLES	
a. a favorable benefit-cost ratio	g. Fig. 23-4c
b. Fig. 23-10	h. fossil fuel industry
c. Fig. 23-4a	i. Sierra Club
d. command-and-control strategy	j. Fig. 23-6
e. trading of emissions allowances	k. Table 23-2
f. Fig. 23-4b	l. cost of air pollution

SELF-TEST

1. Order (1 to 4) the stages of the policy life cycle.

   [ ] implementation [ ] recognition [ ] control [ ] formulation

2. What are the two costs in development of environmental policies?

   a. ______________________________ b. ______________________________

3. Indicate whether the following statements are true [+] or false [-].

   [ ] A strong set of environmental policies means diminished national wealth.
   [ ] A pure form of the free market economy is found in the United States.
   [ ] The dominant source of wealth for most nations is human resources.
   [ ] The calculation of GNP for most, if not all, countries is a conservative estimate.
   [ ] The regulatory approach practically guarantees a certain sustained level of pollution.

[ ] Protecting the environment depends on the strength of environmental interest groups.
[ ] The ultimate responsibility for environmental policy should reside with the public.

4. List two areas of emphasis in environmental policy development.

a. ______________________________ b. ______________________________

5. Using Fig. 23-10, explain the cost of pollution control over time given the percent of pollution reduction.

____

6. At what level (percentage) of pollution control are benefits maximized? ______________

7. How does Fig. 23-10 demonstrate that 100 percent reduction in pollution is not cost-effective?

____

8. Identify the area between the two curves in Fig. 23-10 that represents the area of optimum cost-effectiveness.

____

Indicate whether the following benefits from reduction or prevention of pollution can be assigned a monetary value through [a] real dollar estimates or [b] shadow pricing.

9. [ ] improved human health
10. [ ] improved agriculture and forest production
11. [ ] enhanced commercial and/or sport fishing
12. [ ] enhanced recreational opportunities
13. [ ] extended lifetime of materials and cleaning
14. [ ] enhanced real-estate values

15. The costs over time for pollution control

a. are greatest in the short-term and diminish in the long-term
b. are low in the short-term but increase exponentially the longer the control is in effect
c. are minimal at first, then increase to a maximum, then decrease again
d. are maximal at first, then decrease to a minimum, then increase again

16. Using Fig. 23-11, describe what happens to pollution control costs each year after time of initiation. ____

17. At what point in time does Fig. 23-11 demonstrate a favorable cost-benefit ratio? __________ years.

18. Identify the economic system used in socialist countries: ____________________

19. Predict how environmental policy might be generated without public input.

____

20. Expecting future generations to bear the costs of today's pollution is a violation of (inter- or intra-) generational equity.

ANSWERS TO STUDY GUIDE QUESTIONS

1. b, c, c, a, b; 2. false; 3. improvement of human welfare, environmental protection; 4. degraded environments have a direct negative effect on human welfare; 5. I, I, B, D, D; 6. a, b, c, a, a, a, b, b, b, b, b, b; 7. true; 8. ecosystems, natural resources; 9. produced assets, natural capital, human resources; 10. c, a, e, b, d; 11. human resources; 12. health, education, nutrition; 13. does not calculate for depreciation of natural resources; 14. cannot attach a market value to natural assets, natural assets not considered a part of the stock of a nation's wealth; 15. definition and enforcement of human rights, honest legal system and free press, well-developed market economy, functioning communications and transport network; 16. intergenerational; 17. -, -, +, +; 18. c, a, d, a, b, a, c, d, b, c, d, b; 19. b/c, d, b, b/c, a; 20. effectiveness, efficiency, equity; 21. true; 22. false; 23. positive; 24. states and nations with the strictest environmental regulations have the highest rate of job growth and economic reforms; 25. M, R, R; 26. to control the environmental regulatory process on business and industry; 27. benefit-cost; 28. greater; 29. externality; 30. new jobs or better health; lose of jobs or poorer health; 31. true; 32. equipment purchase and loss of old equipment, implementing control strategies; 33. false; 34. early; 35. see Table 23-3; 36. a, a, a, b, c, c; 37. estimating costs of damages that would occur if regulation were not imposed; 38. 100, threshold; 39. greatest; 40. true; 41. high, decrease; 42. true; 43. acid rain, groundwater pollution; 44. true; 45. pollutant produced in one state has greatest effect in another; 46. all +; 47. 3.6, 50; 48. formulate laws, receive appropriations, authorize appropriations to governmental agencies; 49. all true; 50. mobilize constituents, appropriate large cash contributions, discredit evidence of opponents, influence public and media with telephone calls and letters; 51. a, b, b, d, b

---

ANSWERS TO SELF-TEST

1. 3, 1, 4, 2; 2. monetary, political; 3. -, -, +, +, +, +, +; 4. human welfare, environmental protection; 5. costs steadily increase with each % level of reduction and become prohibitive after 75percent; 6. between 50 and 70 percent; 7. costs exceed benefits prior to 100 percent reduction and costs rise rapidly after 75 percent; 8. area of greatest distance between lines representing value of benefits and cost of pollution control; 9 - 14. all a; 15. a; 16. decrease and level off; 17. about 12 years; 18. centrally planned; 19. would be decided by politicians and special interest groups; 20. inter-

VOCABULARY DRILL ANSWERS

Term	Definition	Example s)	Term	Definition	Example(s)
cost-benefit analysis	d	b	natural capital	h	f
cost-effective	f	a	policy life cycle	a	j
external cost	g	l	produced assets	e	c
GNP	k	k	regulatory approach	c	d
human capital	i	g	social capital	j	g
market approach	b	e	special interest groups	l	h, i

CHAPTER 24

SUSTAINABLE COMMUNITIES AND LIFESTYLES

Urban Sprawl

The Origins of Urban Sprawl

1. List three factors that contributed the most to urban sprawl.

   a. ______________________ b. ______________________

   c. ______________________

2. Urban sprawl is a function of (time, distance).

3. Indicate whether the following factors enhanced [+] or had no effect on [-] urban sprawl.

[ ] ability to purchase cars
[ ] commuting time
[ ] Highway Trust Fund
[ ] commuting distance
[ ] population growth
[ ] the lack of planned development
[ ] low rural tax base
[ ] the environmental or social consequences
[ ] Henry Ford
[ ] low-interest mortgages
[ ] Highway Revenue Act of 1956

4. How have average commuting distances and time changed since 1960?

distance ____________________ time ____________________

5. Explain the difference between a suburb and exurb.

____

Environmental Impacts of Urban Sprawl

6. Indicate whether the following consequences of urban sprawl are related to

a. depletion of energy resources
b. air pollution, acid rain, the greenhouse effect, and stratospheric ozone depletion
c. degradation of water resources and water pollution
d. negative impact on recreational and scenic areas and wildlife
e. loss of agricultural land

[ ] loss of 2.5 million acres per year
[ ] highways through parks or along stream valleys
[ ] decreases infiltration over massive areas
[ ] increased numbers and densities of cars
[ ] increased commuting miles
[ ] increased road kills
[ ] threefold increase in per capita oil consumption

Reining in Urban Sprawl: Smart Growth

7. List four strategies for Smart Growth.

a. ____

b. ____

c. ____

d. ____________________

8. Which of the following are [+] and are not [-] characteristics of Smart Growth?

[ ] disassociated homes, workplaces, recreation areas
[ ] planning for car transportation
[ ] protection of sensitive lands
[ ] community integration

9. Explain the significance of the Intermodal Surface Transportation Efficiency Act of 1991 in promoting suburb sustainability.

____

10. What is the intended function of the TEA 21?

____

Urban Blight

Economic and Ethnic Segregation

11. What is the main, negative, societal impact of exurban migration?

____

12. Exurban migration

a. has profound economic and social impacts (True or False)
b. segregates the population along economic and/or ethnic lines.(True or False)

13. Indicate whether the following characteristics do [+] or do not [-] describe the people who are left behind after exurban migration.

[ ] They are the poor, elderly, handicapped, and minority groups.
[ ] They can obtain mortgage loans.
[ ] They reflect a gentrification of society.
[ ] They represent the economically depressed subset of society.

Vicious Cycle of Urban Blight

14. Local governments

a. are responsible for providing societal services (True or False)
b. operate from budgets derived from property taxes (True or False)
c. are separate entities in the central city and suburbs (True or False)

15. Property values are generally (higher, lower) in the central city when compared to suburbs.

16. Property taxes are generally (higher, lower) in the central city when compared to suburbs.

17. An eroding tax base is characteristic of the (suburbs, central City).

18. The quality and quantity of government services is markedly (better, worse) in the central city when compared to the suburbs.

19. Exurban migration causes the central city infrastructure to (deteriorate, improve).

Economic Exclusion of the Inner City

20. (True or False) People will never return to the central city from the suburbs.

21. (True or False) Some programs are being implemented to revitalize the urban core.

22. Indicate whether the following events are likely to increase [+] or decrease [-] as a result of urban decay.

[ ] industrial development in the central city
[ ] job opportunities in the central city
[ ] the tax base of the central city
[ ] the unemployment rate of central city residents
[ ] social unrest and crime rates within the central city
[ ] the level of education offered to and achieved by central city residents
[ ] accessibility of jobs outside the central city by its residents

23. Suburban sprawl is a (sustainable, nonsustainable) system.

What Makes Cities Livable?

24. The most livable cities around the world have which of the following characteristics? (Check all that apply.)

[ ] reduced outward sprawl
[ ] reduced automobile traffic
[ ] improved access by foot or bicycle
[ ] mass transit
[ ] high population density
[ ] heterogeneity of residences and businesses
[ ] people meet people not cars

25. List three changes used by Portland, Oregon, to make this city more livable.

a. ____________________

b. ____________________

c. ____________________

Moving Toward Sustainable Communities

26. List two outcomes of National Environmental Action Plans.

a. ______________________ b. ______________________

Sustainable Cities

27. (True or False) Cities can be sustainable.

28. Which of the following are characteristics of a sustainable city? (Check all that apply.)

[ ] proximity of people to residences, shops, and workplaces
[ ] need for cars
[ ] use of solar energy
[ ] ability to utilize alternative energy resources
[ ] self-sufficiency in provision of food
[ ] stable population
[ ] cluster development

Sustainable Communities

29. Which of the following were outcomes of the Chattanooga Venture? (Check all that apply.)

[ ] renovation and recycling
[ ] greenways development
[ ] modernized parking garages
[ ] reclaimed waterways
[ ] new industries

President's Council on Sustainable Development

30. What is the purpose of the President's Council on Sustainable Development?

____

31. Economic growth + environmental protection + social equity = ______________ development.

32. Indicate whether the following statements are [+] or are not [-] Council recommendations.

[ ] intergenerational equity
[ ] national goals for achieving sustainable development
[ ] maintained levels of conflict between business and environmental groups
[ ] individual responsibility

Epilogue

33. Define or describe the three themes of how we must live on our planet.

a. sustainability: ______________________________

b. stewardship: ______________________________

c. sound science: ______________________________

Our Dilemma

34. (True or False) All environmental problems can be addressed successfully.

35. (True or False) The root of all environmental problems are humans.

36. List the basic values of stewardship.

a. ______________ b. ______________ c. ______________

d. ______________ e. ______________ f. ______________

Lifestyle Changes

37. Match the actions below with a level of personal change (lifestlye change, volunteerism, political involvement, career choice) required to achieve a sustainable society.

a. recycle: ______________

b. be a joiner: ______________

c. do something for nothing: ______________

d. become something: ______________

VOCABULARY DRILL

Directions: Match each term with the list of definitions and examples given in the tables below. Term definitions and examples might be used more than once.

Term	Definition	Example(s)	Term	Definition	Example(s)
economic exclusion			gentrification		
empowerment zones			urban blight		
eroding tax base			urban decay		
exurban migration			urban sprawl		
exurbs					

VOCABULARY DEFINITIONS
a. residential areas, shopping malls, and other facilities laced together by multilane highways
b. relocation of people, residences, shopping areas, and workplaces in the city to outlying areas
c. communities farther from cities than suburbs
d. segregation of the population into groups sharing common economic, social, and cultural backgrounds
e. declining tax revenue resulting from declining property values
f. whole process of exurban migration and debilitating economic and social impacts on the inner city
g. lack of access by inner city residents to suburban jobs
h. coordinated efforts by government and business to revitalize designated inner city zones

VOCABULARY EXAMPLES		
a. 50 percent unemployment rates	d. suburbia	f. Fig. 24-9
b. Fig. 24-7	e. Fig. 24-8	g. Detroit
c. Fig. 24-3		

SELF-TEST

1. The prime factor that initiated suburban growth and has supported urban sprawl is

   a. farmers selling land to developers.
   b. widespread ownership of private cars.
   c. decline of the central city.
   d. developers buying farms.

2. Which of the following factors had no effect on the rate of urban sprawl?

   a. ability to purchase cars b. commuting time c. commuting distance d. rural tax base

3. An environmental consequence of urban sprawl was

   a. depletion of energy resources.
   b. air pollution, acid rain, and the greenhouse effect.
   c. loss of agricultural land.
   d. all of the above.

4. Which of the following is not a characteristic of the people left behind after exurban migration?

   a. They are the poor, elderly, handicapped, and minority groups.
   b. They have high levels of education and income.
   c. They are predominantly economically depressed minorities and whites.
   d. They reflect the gentrification of society.

5. The factor that led most directly to a decline of the central cities was

a. the most affluent people moved out
b. the rural poor moved into the cities
c. tax revenues of the cities declined
d. the welfare costs of the city increased

6. Indicate whether the following events increased [+] or decreased [-] as a result of exurban migration.

[ ] industrial development in the central city
[ ] job opportunities in the central city
[ ] social unrest and crime rates within the central city
[ ] accessibility of jobs outside the central city to its residents
[ ] level of education offered to and achieved by central city residents
[ ] types and efficiency of local government services
[ ] property tax base of the central city
[ ] unemployment rate of central city residents
[ ] urban decay
[ ] gentrification
[ ] urban sprawl

Use the terms MORE or LESS to complete the following statement.

More urban sprawl leads to 7. __________ exurban migration, which leads to 8. __________ urban decay, which leads to 9. __________ urban sprawl.

The Highway Trust Fund

10. benefits both central city and suburb residents (True or False)
11. brings people closer together (True or False)
12. promotes the establishment of parks and other natural areas (True or False)
13. tends to disrupt existing communities (True or False)

14. One effective way to stem urban blight is to

a. provide mechanisms that give residents ownership in their homes and community
b. give residents large sums of money to purchase goods and services
c. have volunteers come in and fix things up for residents
d. use government funds to build large multiple housing units

15. List one program that gives central city residents ownership of their homes and community.

____

ANSWERS TO STUDY GUIDE QUESTIONS

1. car ownership, low home mortgages, highways; 2. time; 3. +, +, +, -, -, +, +, -, +, +, +; 4. doubled, no change; 5. suburbs were developed as an escape, by some, from the inner city, but exurbs (e.g., rural areas) represent an escape, by some, from the suburbs; 6. e, d, c, b, a, d, a; 7. setting boundaries on urban sprawl, saving open space, developing existing urban space, creating new towns; 8. -, -, +, +; 9. provides funding from highway trust fund to construct modes of access for alternative forms of transportation; 10. provides transportation enhancements that include pedestrian and bicycle facilities; 11. urban decay; 12. both true; 13. +, -, +, +; 14. all true; 15. lower; 16. higher; 17. central city; 18. worse; 19. deteriorate; 20. false; 21. true; 22. -, -, -, +, +, -, -; 23. nonsustainable; 24. all checked; 25. reduced outward sprawl, reduced automobile traffic, mass transit; 26. Sustainable Cities Program, U.S. Council on Sustainable Development; 27. true; 28. check, blank, all the rest checked; 29. check, check, blank, check, check; 30. guide the country toward greater sustainability; 31. sustainable; 32. +, +, -, +; 33. practical interaction between human and natural ecosystems, ethical and moral framework guiding our actions, basis for our understanding of how the world works; 34. true; 35. true; 36. compassion, concern for justice, honesty, frugality, humility, neighborliness; 37. lifestyle change, political involvement, volunteerism, career choice

---

ANSWERS TO SELF-TEST

1. b; 2. c; 3. d; 4. b; 5. c; 6. -; 7. -; 8. +; 9. -; 10. -; 11. -; 12. -; 13. +; 14. +; 15. +; 16. +; 17. more; 18. more; 19. more; 20. false; 21. false; 22. false; 23. true; 24. a; 25. empowerment zones or Habitat for Humanity

VOCABULARY DRILL ANSWERS

Term	Definition	Example(s)	Term	Definition	Example(s)
economic exclusion	g	a	gentrification	d	d
empowerment zones	h	g	urban blight	f	f
eroding tax base	e	e	urban decay	f	f
exurban migration	b	b	urban sprawl	a	c
exurbs	c	f			